SPACE SCIENCE, EXPLORATION AND POLICIES

SPACE TECHNOLOGY CONTRIBUTION IN ENGINEERING SERVICES

SPACE SCIENCE, EXPLORATION AND POLICIES

Additional books in this series can be found on Nova's website
under the Series tab.

Additional e-books in this series can be found on Nova's website
under the e-book tab.

SPACE SCIENCE, EXPLORATION AND POLICIES

SPACE TECHNOLOGY CONTRIBUTION IN ENGINEERING SERVICES

RUSTAM B. RUSTAMOV
BAHAR N. ALIYEVA
NARGIZ A. BABAYEVA
SABINA N. HASANOVA
AND
MARAL H. ZEYNALOVA

nova publishers
New York

Library of Congress Cataloging-in-Publication Data

Space technology contribution in engineering services / editors, Rustam B. Rustamov, Bahar N. Aliyeva, Nargiz A. Babayeva, Sabina N. Hasanova, and Maral H. Zeynalova (Institute of Physics, Azerbaijan National Academy of Science, Baku, Azerbaijan, and others).
pages cm
Includes index.
ISBN 978-1-63483-033-1 (hardcover)
1. Astronautics--Technology transfer. 2. Engineering--Technological innovations. 3. Remote sensing--Industrial applications. I. Rustamov, Rustam B.
TL865.S66 2015
338.9'26--dc23

2015018688

Published by Nova Science Publishers, Inc. † New York

CONTENTS

PREFACE

Remote Sensing is the method of measurement of the object properties on the Earth's surface using data acquired from aircraft and satellites. This method makes it possible to carry out the measurement from the distance for the purpose of this research and to display those measurements over a two-dimensional spatial grid, i.e., images. Remote Sensing method, which is based on the data accessed from satellites, provides an opportunity to observe and monitor the Earth, collect appropriate information with further processing and integration of the Geographic Information System.

This book's aim is to reveal the opportunities of the application of Remote Sensing and Geographic Information System advances in wide areas of engineering applications through analyzing own achievements and challenges as well as creative sources existing in the indicated area.

It is stated as a general approach of using the Remote Sensing and Geographic Information System technology. It includes application of methodologies of the space technology in a various areas of engineering, space image sources. The fact is that the management of the processing stage of space data depends on an engineering task solving, space technology applications in engineering/management, civil/town planning and architectural design, space technology contribution in environmental spheres.

This book performes the cognetive conceptual aspects of Remote Sensing and Geographic Information System widely. Its applications are based on the implemented projects. The outcomes of the offered high technology success found a perfect reliable place in the engineering as an excellent tool.

Keywords: Remote Sensing, Geographic Information System, engineering, architecture, space image processing, field measurements

INTRODUCTION

Present technological boom is posing more challenges in addition to traditional engineering achievements, particularly health and safety issues and on-site hazards. These circumstances demand the technological solutions with a rapid change of the needs. There is no doubt that it makes necessary to use the technological advances including the high technology applications.

As a result, it is important to answer to what is required to be done and what way needs to be selected for more challenges and new achievements? How and which high technology advances can be used and applied depending on engineering solutions.

An answer to the questions above can be easily discovered by understanding of an approach of problem solving, which demands the appropriate information and that is the basis of a main challenges in engineering developments. It is obvious that space technology use and application can play a great role and is a very good instrument for meeting expected requirements.

Honestly, up today the advances of space technology are used in a large area of our life. It takes a strong position in different areas of engineering as well. Definitely Remote Sensing is one of the best methods in collection of data from air and satellite platforms. In the meantime, successful use of appropriate software for data processing and integration into Geographic Information System is a key instrument in engineering starting from the first stage of tender package's preparation up to construction.

This book introduces the detailed concepts of methods of Remote Sensing and Geographic Information System in a various areas of engineering. It also addresses the way of the application based on the offered methods and technologies depending on the problem required to be solved in any stages of the engineering executions. As such, it covers the aspects of the space technology advances while providing a solid basis for engaging in concrete research activities.

METHODS

This book introduces the detailed concepts of methods of Remote Sensing and Geographic Information System in a variety of areas of engineering. It also addresses the way of application based on offered methods and technologies depending on the problem required to be solved in any stages of the engineering executions. As such, it covers the aspects of the space technology advances application providing a solid basis for engaging in concrete activities. It is intended to demonstrate not only the methods of Remote Sensing and Geographic Information System advantages but also success or some conceptual matters. It has been reflected on the actual application of suggested methods within actual executed projects.

Chapter 1

REMOTE SENSING AND GEOGRAPHIC INFORMATION SYSTEM TECHNOLOGY

ABSTRACT

There is no doubt that the use of appropriate technology in the Earth studies is one of the important issues and vital aspects of successful outcomes in the area. In this regard, it is necessary to indicate that Remote Sensing, Geographic Information System and Global Positioning System as a key element of the space technology application provides extremely useful tools for a wide range of problems solving like natural resources studies, Natural disaster, climate change and global human impacts with reflection of consequences, environmental and ecological processes investigations. They are widely recognized as supporting tools for the planning, monitoring, and management of the appropriate utilization of resources on the country, regional and global scales.

Keywords: Remote Sensing, Geographical Information System, global positioning system, electromagnet spectrum, data classification

1.1. INTRODUCTION: A VITAL PLACE OF REMOTE SENSING AND GEOGRAPHIC INFORMATION SYSTEM

The technology mentioned above represents multidisciplinary backgrounds. In this case, Remote Sensing and Geographic Information System shares a common interest, that is, to use remote sensing, Geographic Information System, Global Positioning System and other space technologies as tools in pursuing basic investigation achievements in academic level as well as in developing new technologies that are applicable in the wide areas of problem solving. Because of the complexity of the technologies together with heavy dependence on advanced computer skills, application specialists need to have a sound knowledge of the theoretical aspects and practical approaches to integrate many resources of information that address different applications. Furthermore, scientists, planners, or engineers interested in these technologies should be familiar with past, present and future satellite systems, their appropriate usage, data acquisition and handling and integration with other data sources. For this reason, contribution provided in such sensitive and valuable area is highly important and plays a significant role in understanding and application of purposes [1].

It is important to indicate that the demand for Remote Sensing and Geographic Information System use is very high due to the large number of opportunities of the application of the technology advances. Employment opportunities are available in a wide range of areas, including agriculture, forestry, coastal development and management, urban planning and development, medical technology, mapping and planning, disaster mitigation and environmental management etc.

Furthermore, the areas of Remote Sensing and Geographic Information System needs to be covered in the stage of data processing of such aspects as the Earth-energy interaction, atmospheric correction, application of potential in various disciplines, Geographic Information System data sources, map projection, geo-statistics, spatial modelling, automated mapping, digital terrain model, Global Positioning System data acquisition, the integration of Geographic Information System, Remote Sensing and Global Positioning System.

1.2. Space Science and Technology in Earth Study

Remote sensing is the method of the studying of object properties on the Earth's surface using the data acquired from an aircraft and satellites. It makes available to measure features on the Earth surface at a distance, rather than in situ and for this research's purposes, displays those measurements over a two-dimensional spatial grid, i.e., images accessed from the distance. Remote-sensing systems, particularly those deployed on satellites, provide a repetitive and consistent view of the Earth facilitating the ability to monitor the Earth system and the effects of human activities on the Earth. There are many electromagnetic (EM) band-length ranges that the Earth's atmosphere absorbs. The electromagnetic band ranges transmittable through the Earth's atmosphere sometimes referred to as atmospheric windows.

As per known, the human eye only detects, visible reflective solar radiance which human actually see. It relates to the electromagnetic scale in the band length range 0.4 – 0.7 μm. An advantage of Remote Sensing method allows the detection of other reflective and radiant (e.g., thermal) energy band-length ranges that reach or are emitted by the Earth's surface and even some Earth's atmosphere reflects, e.g., the electromagnetic reflective qualities of clouds. Hence, for viewing purposes red, green and blue (RGB) false colour assignments are used to express the reflective qualities of objects in these electromagnetic band-length groups and the combination and mixing of these false colour assignments express the true physical reflective qualities of all objects presented in an image [2].

Remote Sensing method plays a significant role in a data collection from the distance. It is necessary to indicate that there is one more technology opportunity for more successful use of Remote Sensing data for further useful application. It is a development of Geographic Information System technology. The primary benefit of Geographic Information System is the ability to interrelate spatially multiple types of information assembled from a range of sources. These data do not necessarily have to be visual. Shape files are helpful for interpolating and visualizing many other types of data, e.g., demographic data. Many study and research models rely on the ability to analyse and extract information from images by using a variety of computer available research tools and then express these findings as part of a project with images in a variety of layers and scenes.

In application of satellite images with further processing to assess most types of land cover change, primarily those involving change in vegetation coverage, variations in climate must be considered. For better control, monitoring of the process and accuracy of these analyses, comparing images acquired during the same month or season is advisable. However, due to the limited availability of satellite images, obtaining materials corresponding both spatially and temporarily to the location and period under research is not always possible. Furthermore, annual and seasonal climate data are not always available for the region or temporal period that is researched. Sometimes, changes in average rainfall, temperature, etc. must be inferred using more macro regional or global data. As we can see, there are existing limits in application stages of space technology advances.

A benefit of Remote Sensing and Geographic Information System technology is to study of Land Cover/Land Use properties of the Earth surface. One standard of Remote Sensing application for detecting temporal change in land cover, especially vegetation, is the Normalized Difference Vegetation Index (NDVI). The NDVI application involves a ratio formula between the visual red and near infrared (NIR) electromagnetic bands. This ratio application helps to distinguish healthy and stronger vegetation reflection from other materials with similar reflective qualities in those electromagnetic band wavelength groups. NDVI applications are useful because two images can be processed into a false colour composite, which allows the visual, temporal change detection in vegetation coverage. Moreover, by applying standardized thresholds to multiple NDVI manipulated images, one can create classification-training regions and execute supervised computer-generated classifications of multiple images. From these resulting images, area summary reports are calculated. These empirical data enable a more accurate assessment of change in area of the corresponding land-cover classes.

Information pertaining to some of the above topics, as well as a more comprehensive description on some Remote Sensing technologies including a glossary of terms, have been given in the sections below [3-5].

The Electromagnetic Spectrum:

Gamma rays < 0.30 nm	this range is completely absorbed by the upper atmosphere and not available for Remote Sensing.
X-rays 0.03—30.0 nm	this range is completely absorbed by the atmosphere and not employed in Remote Sensing.
UV-rays 0.03—0.40 µm	this range is completely absorbed by the atmosphere and not employed in Remote Sensing.
Photographic UV 0.30—0.40 µm	this range is not absorbed by the atmosphere and detectable with film and photo detectors but with severe atmospheric scattering.
Visual Blue 0.45—0.52 µm	Because of water increasingly absorbs electromagnetic (EM) radiation at longer wavelengths; band 1 provides the best data for mapping depth-detail of water-covered areas. It is also used for soil-vegetation discrimination, forest mapping, and distinguishing cultural features.

Visual Green 0.50—0.60 µm	The blue-green region of the spectrum corresponds to the chlorophyll absorption of healthy vegetation and is useful for mapping details such as depth or sediment in water bodies. Cultural features such as roads and buildings also show up well in this band.
Visual Red 0.60—0.70 µm	Chlorophyll absorbs these wavelengths in healthy vegetation. Hence, this band is useful for distinguishing plant species, as well as soil and geologic boundaries.
Near IR 0.70—0.80 µm	The near IR corresponds to the region of the EM spectrum, which is especially sensitive to varying vegetation biomass. It also emphasizes soil-crop and land-water boundaries.
Near IR 0.80—1.10 µm	The second near IR band is used for vegetation discrimination, penetrating haze, and water-land boundaries.
Mid-IR 1.55—1.74 µm	This region is sensitive to plant water content, which is a useful measure in studies of vegetation health. This band is also used for distinguishing clouds, snow, and ice.
Mid IR 2.08—2.35 µm	This region is used for mapping geologic formations and soil boundaries. It is also responsive to plant and soil moisture content.
Mid-IR 3.55—3.93 µm	A thermal band which detects both reflected sunlight and earth--emitted radiation as well as is useful for snow-ice discrimination and forest fire detection.
Thermal IR 10.40—12.50 µm	This region of the spectrum is dominated completely by radiation emitted by the Earth and helps to account the effects of atmospheric absorption, scattering, and emission. It is useful for crop stress detection, heat intensity, insecticide applications, thermal pollution, and geothermal mapping. This channel is commonly used in water surface temperature measurements.
Microwave-Radar 0.10—100 cm	Microwaves can penetrate clouds, fog, and rain. Images can be acquired in the active or passive mode. Radar is the active form of microwave Remote Sensing. Radar images are acquired at various wavelength bands.
TV & Radio >10 m	The longest-wavelength portion of the electromagnetic spectrum.

1.3. Satellite Sensors Applied for Remote Sensing Data

There are satellite sensors used for detection of the radiation reflected by the Earth surface:

1.3.1. Advanced Very High Resolution Radiometer

The Advanced Very High Resolution Radiometer (AVHRR) produces 1 km multispectral data from the NOAA satellite series (1979 to present). The AVHRR's four or five spectral bands are used primarily for mapping large areas, especially when good temporal resolution is required. Applications includes snow cover and vegetation mapping; flood, wild fire, dust and sandstorm monitoring; regional soil moisture analysis; and various large-scale geologic applications [5, 6].

Spatial Resolution:	1 km
Spectral Bands:	**Band 1:** (visible red, 0.58-0.68μm)
	Band 2: (near IR, 0.725-1.10μm)
	Band 3: (IR, 3.55-3.93μm)
	Band 4: (thermal IR, 10.30-11.30μm)
	Band 5: (thermal IR, 11.50-12.50μm)

Landsat MSS

The Landsat Multi-Spectral Scanner flew on the first five Landsat missions, providing continuous, comparable data over a period of about 20 years, from 1972 to 1993.

Spatial Resolution:	57 m
Spectral Bands:	**Band 1**: (visual green, 0.50-0.60μm)
	Band 2: (visual red, 0.60-0.70μm)
	Band 3: (near IR, 0.70-0.80μm)
	Band 4: (near IR, 0.80-1.10μm)

Landsat TM

The Landsat TM missions began in 1982 with Landsat-4 and have continued to the present with the Landsat-7 mission.

Spatial Resolution:	30 m
Spectral Bands:	**Band 1**: (visual blue, 0.45-0.52μm)
	Band 2: (green, 0.52-0.60μm)
	Band 3: (red, 0.63-0.69μm)
	Band 4: (near IR, 0.76-0.90μm)
	Band 5: (mid IR, 1.55-1.74μm)
	Band 6: (thermal IR 10.40-12.50μm)
	Band 7: (mid IR, 2.08-2.35μm)

Landsat ETM+

The Enhanced Thematic Mapper Plus (ETM+) sensors record data using the same seven bands as the TM sensors. One advanced feature of this enhanced sensor is the addition of a panchromatic band with 15 m spatial resolutions and a bandwidth from 0.52 to 0.90 μm. The second major enhancement is the increase in spatial resolution of the thermal band (6) from 100 to 60 m. It was launched in 1999 on the Landsat-7 mission.

Glossary:

Albedo	Ratio of the amount of electromagnetic energy (solar radiation) reflected by a surface to the amount of energy incident upon the surface.
ASTER	Advanced Space borne Thermal Emission and Reflection Radiometer.
AVHRR	Advanced very high-resolution radiometer.
AVIRIS	Airborne visible-infrared imaging spectrometer.
Band	Broadcasting frequency within given limits. A subdivision within an electromagnetic region.
Bandwidth	The total range of frequency required to pass a specific modulated (*spectral resolution*) signal without distortion or loss of data. The wavelength interval recorded by a detector.
CEO	Centre for Observing the Earth from Space at Yale University
ETM+	Enhanced Thematic Mapper Plus
EM	Electromagnetic
GPS	Global Positioning System
GIS	Global Information System
IFOV	Instantaneous field of view: the solid angle through which a detector is sensitive to radiation. In a scanning system, the solid angle subtended by the detector when the scanning motion is stopped.
IKONOS	A high-resolution Earth observation satellite launched in 1999, which occupies a 682-km sun synchronous orbit and employs linear array technology collecting data in four multispectral bands at a nominal resolution of 4 m, as well as a 1-m-resolution panchromatic band.
Landsat	A series of unmanned NASA Earth resource satellites that acquire multispectral images in the visible and IR bands.
NAD	North Atlantic Datum
NDVI	Normalized Difference Vegetation Index
NIR	Near Infrared Remote Sensing of energy naturally reflected or radiated from the terrain.
Radiation	Act of giving off electromagnetic energy.
RGB	Red, Green, and Blue—the colours used in constructing visible and false colour image representations.
MIR	Mid Infrared
Spatial Resolution	The ability to distinguish between closely spaced objects on an image. Commonly expressed as the most closely spaced line-pairs per unit distance distinguishable.
Spectral Reflectance	Reflectance of electromagnetic energy at specified wavelength intervals.
Spectral Resolution	Range of wavelengths recorded by a detector. *(bandwidth)*
SWIR	Short Wave Infrared
TM	Thematic Mapper

UTM	Universal Transverse Mercator
VI	Vegetation Index
VNIR	Visible Near Infrared
WGS	Worldwide Geographic System
WRS	Worldwide Reference System

1.4. GEOGRAPHIC INFORMATION SYSTEM

Geographic Information System is one of the main instruments in space science and technology for data collection and processing contenting classifications of features in the Earth surface. It is comprehensible that, the used word 'Geographic' deals with the spatial objects or features, which may be referenced or related to a specific location on the Earth surface. The object may be physical / natural or may be cultural / manmade. Likewise, the word 'Information' deals with the large volume of data about a particular object on the Earth surface. The data includes a set of qualitative and quantitative aspects, which the real world objects acquire. The term 'System' is used to represent system approach where the complex environment (consists of a large number, of objects / features on the Earth surface and their complex characteristics) is broken down into their component parts for easy understanding and handling, but is considered to form an integrated whole for managing and decision-making. Nowadays this is possible in a very short span of time with the development of sophisticated computer hardware and software. Therefore, Geographic Information System is a computer based information system, which attaches a variety of qualities and characteristics to geographical location and helps in planning and decision-making [7].

A Geographic Information System can be defined in different manners. For example, in some cases, Geographic Information System as a computerized system facilitates the phases of data entry, data analysis and data presentation especially in cases when we are dealing with geo referenced data. At the same time, there is another approach of definition of Geographic Information System as a system, which provides a computerized mechanism for integrating various geoinformation data sets and analysing them in order to generate information relevant to planning needs in a context. According to the description, the Geographic Information System is reflecting as a computer based tool for mapping and analysing things that exist and events that happen on the Earth. Considering current place of Geographic Information System application, it is a set of tools for collecting, storing, transforming, and displaying spatial data from the real world for a particular set of purpose. A Geographic Information System as a computer based system provides five sets of capabilities such as to handle geo referenced data, data input, data management (data storage and retrieval), manipulation analysis and data output.

From the above definitions, it can be concluded that a Geographic Information System user expects support from the system to enter geo-referenced data to analyse it in various ways and to produce output (maps and other) from the data. A Geographic Information System draws on concepts and ideas from many different disciplines, such as cartography, cognitive science, computer science, engineering, environmental sciences, geodesy, landscape architecture, law, photogrammetry, public policy, Remote Sensing, statistics and surveying. So, it involves not only the study of the fundamental issues arising from the creation, handling, storage and use of geographic information, but also examines the impacts of Geographic Information System on the individuals and society and the influences of society on Geographic Information System.

Keeping long tradition of map-making as a background, Geographic Information System has been developed during mid-20th century with the development of computer science. Computers in government organizations and universities in USA were doing the data analysis of geographic locations during 1950s and 1960s. Dr. Roger Tomlinson, Department of Forestry, and Rural Development, Canada developed the first true operational Geographic Information System. It was called as Canada Geographic Information System (CGIS) and used to store analyse and manipulate land related data. Dr. Roger Tomlison was also known as the "Father of Geographic Information System." In 1964, Howard T. Fisher established a laboratory of Computer Graphics and Spatial Analysis at the Harvard Graduate School of Design. This organization developed a number of important theoretical concepts of spatial data handling and in 1970s it distributed seminal software code and system such as 'SYMAP,' 'GRID' and 'ODYSSEY.' This inspired subsequent commercial development. By early 1980s, M & S Computing (later Intergraph) and Environmental Systems Research Institute (ESRI) emerged as commercial vendors of Geographic Information System software. ESRI released ARC/ Info and ARC View software in 1981 and 1992 respectively. By the end of 20th century, the development of ARC View enabled viewing Geographic Information System data through internet and eliminated many of the hardware and licensing expenses of software packages. Since then a number of organizations and universities have been doing research in the field of Geographic Information System and developing user-friendly software. Now there are a growing number of free, open source Geographic Information System packages, which run in a wide range of operating systems and perform specific tasks [1].

1.4.1. Development Stages of Geographic Information System

Content of Geographic Information System primarily integrates geographic data to be analysed, manipulated, and managed in an organized manner through computers to solve real world problems. It is obvious that for development of Geographic Information System is required two main items – computer system (a) and geographic data (b).

(a) Computer System includes both hardware and software. A Geographic Information System runs through computer system ranging from portable personal computers (PCs) to multi-user super computers, programmed by wide variety of software languages.

In all ranges, there are number of things that are essential for effective Geographic Information System operation. These include:

1. a processor with sufficient power to run the software;
2. sufficient memory for the storage of large volume of data;
3. a good quality, high resolution colour graphic screen; and
4. Data input and output devices (for example digitizers, scanners, keyboard, printers and plotters).

There are a wide range of software packages for Geographic Information System analysis, each with its own advantages and disadvantages. Even those lists are too long to be mentioned here, the important ones are different versions of ARC View, ARC Info, Map Info., ARC GIS, Auto Cad Map etc.

(b) Data for Geographic Information System: A Geographic Information System without data is like a car without fuel. Without fuel, a car cannot move, likewise without data a Geographic Information System will not produce anything. Data for Geographic Information System can be obtained from different sources like aerial photographs, satellite imageries, digital data, conventional maps, Census, Meteorological department, field data (surveys/GPS) etc. These data obtained from various sources can be classified into two types – spatial data that describes location and attribute data, which specifies the characteristics at that location. Spatial data tells us, "Where the object is?" And attribute data tells us "What is the object?" or "How much the object is?" In other words, it tells the characteristics at that location.

Spatial data or real world features are very complex. So, spatial data is simplified before they are entered into the computer. The common way of doing this is to break down all geographic features into three basic entity types – points, lines, and areas. Points are 'one dimensional' objects, used to represent features that are very small, e.g., a post box, an electric pole, a well or tube well etc. Only latitudinal and longitudinal values or a coordinate reference can be given to these features to explain their location. Lines are two-dimensional objects and are used to represent linear features, for example roads and rivers. Lines are also used to represent linear features that do not exist in reality, such as administrative boundaries and international boundaries. Areas are three-dimensional objects, represented by closed set of lines, and are used to define features such as agricultural fields, forest areas, administrative areas etc. Area entities often refer to as polygons.

The representation of real world features using the point, line, and area entity types appears relatively simple. However, the appropriate entity to represent real world features is often difficult and it depends upon the scale of the map. On a world map, points represent cities. It only gives information about number of cities shown on the world map. At national or regional scale, the 'point' entity to represent cities is considered too simple, as it tells us nothing about the real size of the city. In this case, cities are represented by 'area' entity. At the local scale, 'area' entity to represent cities would be considered too simple. In this case, cities are represented by mixture of 'point,' 'lines' and 'areas' as entities. Points may be used to represent features such as electric poles, post boxes etc. Likewise, lines and areas may be used to represent road networks and residential blocks respectively. Therefore, the decision makers decide the 'entities' through which different features of real world would be represented.

Attribute Data reflects characteristics of different objects / features on the Earth surface. These are descriptions, measurements, or classification of geographic features. Attribute data can be both qualitative (like land use type, soil type, name of the city/river etc.) and quantitative (like elevation, temperature, pressure of a particular place, crop yield per acre etc.). Therefore, the attribute can be both numeric and textual.

1.4.2. Geographic Information System Operation

The operation of Geographic Information System becomes clear if there is enough experience and knowledge about computer system and different types of data (spatial and attribute) exist. The stages of Geographic Information System operations are reflected below:

i. Data Entry

Both spatial and attribute data are entered into computer system by different input devices like scanner, digitizer, key board, mouse etc. Scanner, digitizer, mouse are used for entering spatial data. The attribute data available as reports, tables etc. are entered through keyboard. As the data is drawn from different sources, they have different scales, projections, referencing system etc. Therefore, there is a need to standardize the database. A Geographic Information System software enables this operation by 'geo referencing' or 'rubber sheeting' method. It means to stretch the maps in different directions for matching with known coordinates.

ii. Storing of Data

The different spatial entities, which represent different features of the real world, can be stored in two different formats in the computer – in raster format and in vector format. The knowledge of these formats in which spatial data are stored, is required for decision makers as it affects the accuracy of data, their analysis, the storing capacity of the computer etc.

In ***raster format*** the entire study area is divided into regular grid or square cells organized into rows and columns. Individual cells are used for storing point, line, and area entities. The point data are stored in individual cells. The line data are stored by linking cells into lines. The area units are stored by grouping cells into polygons.

In ***vector format*** points, lines, and area entities are stored by using coordinate system. Single (x, y) coordinate pair stores a point. A line entity is stored by connecting a series of points into chains. Likewise, areal entity is stored by connecting a series of points into polygons. The number of points used to show spatial features is important in vector format, because it influences the accuracy of spatial features.

The raster and vector format have their own advantages and disadvantages. Raster data led themselves for use with Remotely Sensed data, because digital Remote Sensing data are already presented in raster format. Compared with the raster Geographic Information System, some of the vector Geographic Information System operations are more accurate. For example, the estimates of area based on polygons are more accurate than counts of raster cells, because raster cell boundary is not accurate and contains a part of an adjacent feature. Vector format allows complex data to be stored in a minimum space. Though the raster data is voluminous, it is more amendable for many spatial analyses. Nowadays the advantages can be taken from both the formats by conversing raster data to vector data and vice versa.

The attribute is stored in table form as shown in Table 1, 2 and 3. Then they are linked to their respective spatial data. For raster data, the attribute data of a particular feature is linked to cells forming that feature. For vector data, the attributes are linked to the point, line, or polygon as the case may be. There is a number of ways in which the attribute data can be organized. The relational database model is the most popular. Consider Table 1, which gives information about different tube wells. The entire table is called a 'relation.' Each row has the relation, which corresponds to the information about a given tube well, is called record. Each column, which gives different information of the tube well, is called a 'field' or 'attribute.'

Table 1. Attribute Data of 'Point' Entity

(Tube Wells) Tube Well code	Year of Construction	Depth of water table (m)	Chlorine content of water (mg/ litre)	Water extracted per day (litre)
1	1991	20	150	1500
2	1985	10	120	1600
3	2000	25	200	1450
.				
.				
.				
.				
N	2002	15	100	2000

Table 2. Attribute Data of 'Line' Entity (Roads)

Road Code	Constructed in year	Type of Road	No. of buses plying / day	Heavy traffic hour
1	1981	Metalled	500	8 am – 11 am
2	1985	Metalled	550	9 am – 12 noon
3	1977	Unmetalled	200	7 am – 8 am
.				
.				
.				
.				
N	1975	Unmetalled	150	8 am – 9 am

Table 3. Attribute Data of 'Area' Entity (Forest)

Code of Forest Patches	Tree Density	Type of Species	Occurrence of Forest Fires (Since 1990)	No. of Tribal Living
1	60	Sal	50	60
2	70	Mangrove	65	50
3	25	Bamboo	70	45
.				
.				
.				
.				
N	45	mixed	25	70

1.5. Remote Sensing and Geographic Information System Technology Applications

Today space science and technology takes valuable position in huge areas of applications. It is not only the source of the use of space technology outcomes in selected spheres identified but also founded out in the beginning stage of its applications. The main advantages of Remote Sensing with further Geographic Information System development are to bring parties together, spread and improve idea, support the decision-making process and evaluate projects for the final stage of the execution [8].

The future of Remote Sensing and Geographic Information System technologies will be interoperable. As such, it offers opportunities to improve and structure the information about the planning process and to make it accessible for all participants in the planning process and the public. The general options are to confine a search for new techniques for dealing with a large-scale areas and applications of Remote Sensing and Geographic Information System technologies at various stages of planning, implementation and monitoring of the projects execution.

It is obvious that Remote Sensing and Geographic Information System technologies developed as separated disciplines for both cultural and information technology reasons. It is almost 30 years of essentially separated development that Remote Sensing application begins to recognize the advantages of integrating Geographic Information System technology dataset in to image analysis. It was classified that Remote Sensing is a measurement and analysis of electromagnetic radiation reflected from, transmitted through, or absorbed and scattered by the atmosphere, the hydrosphere and by material at or near the land surface, for the purpose of understanding and managing the Earth's resource and environment.

A Geographic Information System technology classification is an integrated system of computer hardware, software, and trained personnel linking topographic, utility, facility, image and other resource data that is geographically referenced [9].

In all cases, however, the linkage between Remote Sensing and Geographic Information System is clear and intelligent for preparation of base-map, formulation of planning proposal that acts as monitoring tool during implementation phase. A benefit from merging of Geographic Information System and Remote Sensing technologies is well recognized. On one side, Geographic Information System can be used for improving the potential of information extraction of Remotely Sensed data. On the other side, Remotely Sensed data can be used for updating the information of the Geographic Information System.

1.5.1. Contribution of Remote Sensing and Geographic Information System Technology

The Town Planning can be defined variously as the formulation of alternative patterns of urban settlement, the rational use of resources to alleviate urban problems, and the provision of physical and social infrastructure of the city, transportation, utilities, housing, community facilities, and service. The town planning application using Remote Sensing with further development of Geographic Information System technologies is one of the many areas where such utilities can be used for managing planning activities. The planner can look for different options and choose the best suited for the result. The automated process is not only faster but also can be monitored effectively at any eventuality [10].

There are number of reasons making necessary the use of Remote Sensing and Geographic Information System. With multi-temporal analyses, Remote Sensing gives a unique perspective of how cities evolve in town planning process. The key element for mapping rural to urban land use change is the ability to discriminate between rural uses (framing, pasture forest) and urban use (residential, commercial, and recreational). Remote Sensing method can be employed to classify the types of land use over large areas in a practical, economical, and repetitive fashion.

As an information management tool, Geographic Information System was originally a project-specific tool intended to deal with map-based physical application, such as tax mapping or surveying, inventorying of utility networks running mainframes or minicomputers. Another importance is that it is a fast moving and continuously developing technology. Future Remote Sensing and Geographic Information System software will be interoperable and web based. There will be no room left for specialist, standalone database system and they are moving from 'close' to 'open' computing. This change has been set into motion because the users want their corporate information system to handle any data regardless of the data's complexity or sours, including spatial information.

It is necessary to undertake how Remote Sensing and Geographic Information System technology database can be managing rezoning exercise in a neighbourhood property. In this case the problem Scenario zoning for classification of the selected area for investigation is divided into land use districts. Example of general land uses districts and residential, industrial, commercial, and public open space. These can be classified into more detailed categories based on Geographic Information System technology capacity. Over a period of time, changing population and commercial needs often necessitate the changes to zoning plan. With Remote Sensing and Geographic Information System it is needed to be understood how cities will go to increase and protect what they need to grow.

1.5.2. Sources for Processing

There is no doubt that data access and collection is one of the main requirements in future developments. For this reason, it is important to know what the meaning of image processing and classification is. The intent of the classification process is to categorize all pixels in a digital image into one of the several land cover classes, or "themes." This categorized data may then be used to produce thematic maps of the land cover present in an image. Normally, multispectral data are used to perform the classification indeed [11]. The spectral pattern presented within the data for each pixel is used as the numerical basis for categorization. The requirements for rural/urban change detection and mapping applications are:

(1) High or required spatial resolution to obtain detailed information;
(2) Multispectral optical data to make fine distinction among various land use classes.

The full stages of the process of Remote Sensing methods and Geographic Information System methods can be described as following:

1. Classification of old and new satellite images with Remote Sensing methods use (ER Mapper) with scale 30 x 30 resolution;
2. Comparison between both classified images and Remote Sensing method;
3. Building a sample data base with Geographic Information System technology;
4. Analysis and results.

In some cases of study, it can be used the supervised classification features to transform multispectral image data into user-defined thematic information classes. In the references, case was applied a scale 30x30 for the general information investigation like:

(i) regional land use;
(ii) ecological analysis;
(iii) soil;
(iv) water;
(v) vegetation and built area, etc. required by regional planning and resource management.

This resolution can meet the needs of the requirements that are demonstrated by many case studies and practical projects.

1.5.3. Selection of the Area and Supervised Classification of Data

Selection of the area is very important implementation in application of Remote Sensing and Geographic Information System technology. In the first way to choosc the area, which can be used and is able to make a supervised classification depends on the data required from satellite image. It usually takes a sample or known point/points (bench marks) on the Earth surface or ground to analyse and determine the position of these samples with Global Positioning System. After determining these points on satellite image in such way will give more accurate classification of the features on the Earth. Therefore, an action of merging of the Earth features into space image is the next stage of data processing in classification.

After choosing a region's area it can be made a supervised classification with ER Mapper Remote Sensing tool of satellite image [12]. In the final stage of the processing within current status it can be compared between both satellite images through tables and diagrams or with processed Remote Sensing data.

The last stage of process is Geographic Information System development. Therefore, Geographic Information System contains the monitoring and managing future growth, such as new features, for instance, school for accommodation of the growing number of school children in the selected area and developing streets for accommodation of the growing number of population or increase a residential area. Based on the Geographic Information System development such as Geo Media Pro.4 it is necessary to integrate the last supervised classification of image in order to make a digitizing and build a simple database presenting the determination of a new school, for example, through understanding situation and condition in the highly populated area.

In conclusion, it illustrates the importance of computer-aided tools for day-to-day activities. With this introduction for Remote Sensing and Geographic Information System they should never be standalone operation. The Remote Sensing and Geographic Information System are being developed to support the town planning process and will be a part of a larger system. A modern technology is a component based on information which makes it possible to deliver scalable, continuously available, timely and useful information from a limitless variety of sources to analysis and update.

CONCLUSION

Description in the Chapter 1 about Remote Sensing and Geographic Information System applications are evidencing importance and usefulness of huge areas of studies. It captures the Earth studies participating and supporting the achievements of expectations in large spheres. This Chapter considers the Remote Sensing and Geographic Information Systems applications. It consists of stages of data collection, analyses, integration stages and processing. The product as a result of all executed stages can be performed in the form of maps, tables or any other kind of e-versions.

Chapter 2

METHODOLOGIES OF SPACE TECHNOLOGY APPLICATIONS

ABSTRACT

There are methodologies in space science and technology application used in different areas. This chapter considers Remote Sensing methods and Geographic Information Systems technologies in the Earth study. These methodologies have been undertaken in design of processes for solving wide problems within the engineering stages of executions.

It is important to know what methodology and methods exist. At the same time, the selection and classification of required methodology and methods with the definitions within existing tasks for the satisfaction of needs properly and in time is also important.

One of the significant factors of implementation is correct finding and giving definitions of required information bases that are based on remotely accessed data and field measurements where integration has to be done as the next stage of development.

Keywords: methodology, model development, engineering, and engineering solutions

INTRODUCTION

The modelling of accessed data is required for any process's execution. It relates to the case of Remote Sensing data consisted of large and multidisciplinary purposes in the engineering activities. It can be applied from the first stage of engineering developments. There are undertaking case studies of the data collection and interpretation and the factors needed to be considered in the engineering executions. A study of segments like land Use/Land Cover, soil investigations are the main aspects of a suitable use of Remote Sensing and Geographic Information System as a method of advance technology in the engineering solutions.

2.1. Data Modelling by Use of Remote Sensing and Geographic Information System

The modelling approach of the west water studies by using of the Wastewater Master Plan study is reflected and has been considered within the study of water flow monitoring, developed population and land use projections, developed a hydraulic model of the collection system, and identified existing and future constraints within the collections system.

The study was oriented to the following directions:

- inadequate existing capacity;
- population growth;
- changes in service area;
- Potential reuse projects.

The study was implemented under project and it was discovered that an existing capacity issues in the collection system had caused some sanitary sewer overflows (SSO's). The place selected for the study is primarily interested in eliminating the occurrence of SSO's. Limited capacity in portions of the collection system was identified through the interviews with the city staff and by running the hydraulic model. The model helped to confirm some problem areas and to uncover others. Population growth is not a big driver for the selected area. The total projected population growth over the next 30 years is approximately 10% of the current population. Growth in some areas will be greater than others especially, in areas planned to be annexed by the area. It is in these projected growth and annexation areas that particular attention was given to ensure adequate capacity for the future.

The area was also interested in exploring possible locations for a satellite wastewater treatment facility. The plan was to find out a location where a satellite plant might scalp flows from the collection system treat to reuse levels and then pump into the reuse distribution system. The idea had particular merit, because the largest reuse customers in the city are located on the west side of the town. The current wastewater treatment facility is located on the northeast edge of the city. Because of the existing capacity problems across the system and the need for planned expansion, a hydraulic model of the system was needed. Selection of a predictive hydrologic and hydraulic model for the sewer system is a key element of the master planning process. The selected model must be capable of analysing current loads, determining theoretical hydraulic and hydrologic capacity, and identifying and locating deficiencies in the existing systems. The selected model must also be able to analyse future growth loads imposed on the existing systems and evaluate scenarios for handling these loads.

The selected model should serve as a tool to provide reasonable results for planning efforts and be easily updated for reflecting ongoing land use changes and development. The model should be user friendly and allow the import and export of data for utilization and maintenance by decision makers. Key points of evaluation included:

- integration/interface with Geographic Information System software;
- single software vendor for both water and sewer models;
- cost;

- interface with City data;
- User friendliness.

A survey was conducted on commercially available sanitary sewer collection computer models. The vendors and programs listed in table 4 were compared based on the criteria listed above. After comparison, the MWH Soft product H2OMap SWMM was selected as the best modelling software for the selected area.

Table 4. Sewer Modelling Software Packages Vendor Geographic Information System Integrated Geographic Information System Interface

Vendor	Geographic Information System Integrated	Geographic Information System Interface
MWH Soft, Inc.	InfoSWMM/ InfoSewer	H20Map Sewer/ H20Map SWMM
Haestad Methods/Bentley	SewerGEMS	SewerCAD
Wallingford Software	--	InfoWorks CS
Danish Hydraulics Institute (DHI)	--	MikeURBAN CS

2.2. Model Development

As part of the Wastewater Master Plan project for the City of Abilenc, the project team of Enprotec/Hibbs & Todd (e/HT) and HDR Engineering (HDR) has:

- Collected existing CAD and Geographic Information System based maps for the wastewater collection system;
- Inventoried the existing data to determine information needs;
- Formatted and geo-processed the existing maps and databases for the use in the hydraulic model, H2OMap SWMM.

The area selected for the study has wastewater system data as provided to e/HT and HDR, consisted primarily of two Geographic Information System files: SANMHS (sanitary sewer manhole data) and SANLINE (sanitary sewer line data). Additional data including Geographic Information System files containing parcel information, roads, and railroads were also provided to the staff involved in project execution.

A CAD map of the sanitary sewer system was provided as a supplement to the Geographic Information System files. The SANMHS file included 9,727 features representing manholes within the areas' Wastewater Collection System. A manhole features, 3,088 were missing rim elevation and/or manhole depth information. A list of approximately 200 manholes was created for the field verification of the missing data. These manholes were associated with larger diameter collection system pipe and other critical model elements [1].

The SANLINE file contained 15,883 line features, representing the areas' wastewater collection lines. These features were left without diameter attribute data and were not utilized in the model. Diameters for the remaining 189 features were either surveyed or estimated as accurately as possible or determined not to be critical components of the model and left unchanged.

In the SANMHS file provided, polyline features represented manholes. The modelling software required that the manholes to be represented as point features. First, the polyline features were converted to polygons. For most of the manhole features, a script automated this process. A script is a set of instructions, written in a script language like VBScript which can be used to automate tasks such as data conversion within the Geographic Information System environment. However, for many of the features the polyline start and ends did not match and hand editing was required for conversion of the complete set of manhole data.

After all of the manhole features had been converted to polygons, a point feature was assigned to the centroid of each of the converted polygon features. The new point features were updated with their associated identification numbers, rim elevations, invert depths and other attribute data using a spatial join process.

Formatting of the data is an important factor to consider when developing a Geographic Information System based on the map of any infrastructure system. A particular importance for pipe networks, the pipes must meet at points for most modelling packages to represent the system correctly. Much of the model construction time for this project was spent on formatting the data of the city though it was already in the correct format. Developing the maps in the correct format from the beginning may save considerable time and effort when the maps are subsequently used for the model development.

It has been conducted flow monitoring of the selected area for the purposes of this study. Results of the flow monitoring were used in the determination of flow composition in the area. Flow is generally comprised of three key elements:

i. base wastewater flow;
ii. groundwater infiltration; and
iii. rainfall - dependent infiltration/inflows.

The hydrographs generated from the flow monitoring can be used for estimating these three components for each sub-basin.

Base dry weather flow (BWF) is the flow contributed to the system directly from customers. Flows are generally diurnal, which means they follow patterns throughout the day with peaks generally occurring during the morning between 7:00 am and 9:00 am and again in the evening from 6:00 pm to 10:00 pm. Commercial and industrial flows of the diurnal patterns stay more constant during the workday. Weekdays and weekends typically exhibit different flow patterns; therefore, we used weekday average flows. Base dry weather flows can be broken down further into three basic components: residential, commercial, and industrial flows.

Groundwater infiltration (GWI) is defined as the constant inflow of groundwater into the collection system. This value is determined by several factors including the location of the groundwater table and the physical condition of the collection system. Flow monitoring evaluated during the late night and early morning hours is a general characteristic of GWI in systems that do not have substantial, industrial and/or commercial flows that enter the system throughout the 24-hour period. GWI can also be impacted by dry and wet periods. Rainfall-Dependent Infiltration/Inflow (RDII) is a portion of flow that enters the collection system during a rainfall event. The methodology used to generate the RDII hydrograph was based on a unit hydrograph approach where the characteristics were derived from the flow monitoring data, and were defined as R, T, and K. The R was originally defined for each sub-basin in the flow monitoring report as average leakiness factor. This is a percentage of a rainfall that entered the system within each sub-basin. T is the time of concentration, or the time difference between when the rainfall started and when the peak flow hydrograph was observed at the monitoring locations. K is the recession coefficient, or the ratio between the peak of the flow hydrograph and the point at which the RDII receded.

RDII was found to be a significant portion of total wet weather flows in the system. Infiltration may be the result of many varying issues, including but not limited to poor construction in high groundwater, root intrusion, offset joints, or broken or crushed pipe. Many of the issues addressed can be remedied through trenchless solutions. The following gives a brief description of some of the available rehabilitation techniques in the area selected for study which could be used to address deficiencies.

2.3. Main Replacement

This has become a frequently used technology to replace aging sewer main that may also be experiencing capacity issues. A "bursting head" is routed through existing pipe and a flexible new pipe is pulled from behind. In sewer projects, this can be done from manhole to manhole.

However, it should be noted that services would then need to be handled individually. This is an effective means of replacing sewer main without causing major surface impacts, but will require trenching and replacing of the sewer service. Pipe bursting is particularly effective with vitrified clay pipe. However, low areas or sags cannot be addressed with this method.

Slip Lining – Slip lining sewer main includes installing a rigid sleeve inside the existing sewer main. This is installed from an access point at one end of the main to be rehabilitated.

Cure-In-Place – Similarly, to slip lining techniques, deficiencies involving wide, offset joints should not be slip-lined and low areas are not addressed with this method.

Main and Service Line Replacement - By replacing both the main and service lines that are within the area of the sewer main trench, infiltration can be substantially reduced if proper construction practices are used.

The service lines tend to be shallow and not affected by groundwater until they get close to the mains in areas where basements do not exist. Replacing the portion of the service line within the main trench should target the majority of the service line impacted by groundwater. The disadvantage of this method is a high cost and disturbance of improved areas. This method is estimated to remove approximately 90-percent of infiltration from the sewer main that is rehabilitated. Table 5 lists advantages and disadvantages of these various rehabilitation techniques. Open Cut Replacement is the Most Disruptive Method.

Table 5. Common Rehabilitation Techniques and Advantages/Disadvantages

Method	Applications	Advantages / Disadvantages	Comments
Slip lining	Continuous or segmented pipe with flush joints	Generally cost-effective and timely. Requires excavation for insertion pits and lateral connections, can reduce hydraulic capacity, and usually requires bypass pumping.	Widely used and economical; however can cause disruptions and applications for severely damaged or deteriorated conduits are limited.
Cured-in-Place Pipe (CIPP)	Continuous or segmented pipe with flush joints	Generally cost-effective and timely. Can be installed from existing structures – no excavation required. Very little reduction in hydraulic capacity. No resulting annular space to be grouted. Requires bypass pumping. Applications limited by steep grades.	CIPP uses a flexible fabric liner saturated with a thermosetting resin forced into position with water or air. Applications for severely damaged or deteriorated conduits are somewhat limited.
Pipe-Bursting Replacement	Most pipe materials for which diameter fordiameter or upsizing by one to two diameters is desired	Advantageous when open cut is difficult. Can correct most pipe defects. Improved hydraulic capacity. Typically requires excavation for insertion pits and service connections. More costly than "liner" methods. Requires full or Partial bypass pumping.	Process breaks the old pipe outward and pulls in replacement pipe. Can be cost-effective when excavation is difficult and density of service laterals is low.
Open Cut Replacement	All applications	Repairs all defects and provides the longest service life. Most expensive and disruptive. Requires bypass pumping, traffic control, excavation, backfilling, surface restoration, etc.	Traditional pipe rehabilitation method.
Point Repairs	Small, localized problems	Cost-effective and timely. Limited applications.	Accessed from pipe interior or from open cut. Used typically for isolated "spot" repairs only.

Red text indicates disadvantages.

2.4. Model Results and Recommendations

One of the most important steps in developing recommendations for collection system improvements is the selection of an appropriate design criteria for system sizing. In many cases, a utility may not have a set of pipe sizing design criteria on which to base the proposed system improvements. If that is the case, a set of criteria to be used for system layout for future conditions must be agreed upon by the city staff before the system improvements can be effectively sized.

For this project, the following design criteria was used for sizing system improvements.

- Size pipes for 75% of full flow capacity at the design flows;

- Design flows are based on the measured average residential flow contribution of 78 gal/capita/day;
- Diurnal patterns for future flows are based on the measured flows;
- Design flows for RDII are based on calculated hydrologic values, rather than a fixed input per area served;
- Design flows are based on 2030 wet weather flows developed from a 10-yr, 3-hr storm event.

The design flows were projected out to the 2030-planning horizon using projected population data and the measured average residential wastewater contribution of 78 gal/capita/day. This was the measured system wide average dry weather residential flow component and was the most accurate number available. The daily peaks in the 2030 system were generated by using the same diurnal flow pattern that was observed during the flow-monitoring period.

The RDII contribution to the 2030 wet weather flow design condition was based on hydrologic simulation of a 10-year rainfall event in the model. The leakiness of the system (as determined during the flow monitoring and discussed above) was used to estimate the amount of RDII generated by the design storm event. This method of the estimation of the amount of RDII is more accurate than using a fixed area contribution factor. Some limited reduction in RDII may be assumed to occur as portions of the collection system are replaced or slip lined, but additional leaks will develop to offset these improvements. Therefore, the total RDII contribution for each sub-basin is assumed to remain relatively constant.

Finally, the decision was made to size pipes to accommodate the projected 2030 peak hours wet weather flows with a maximum flow to full pipe capacity ratio of 75%. This value allows for some additional flow to be added to the system due to unforeseen development, additional RDII or greater than expected population increase. This design capacity ratio also allows for some additional pipe conveyance capacity to conservatively design the system since much of the design flow determination is based on estimated values.

2.5. Methodology in Case Study with Remote Sensing and GIS Technology

Land Use/Land Cover studies are the main parts of the area selection in stages of expected implementations. It plays an important role in assessment of land use suitability for any kind of works in urban and rural planning processes. Today, it is very difficult to use traditional methods for planning particularly for urban and rural purposes as well. It makes it impossible impersonally and quantitatively to indicate the difference in land use suitability, and make urban and rural planning more subjectively. Because there are number of reasons having influences on a land use suitability and giving different types of impacts on an objective evaluation of the condition. It limits the development of the expected level of urban and rural planning process. It is necessary to explore a land of urban and rural planning quantitative evaluation method to provide a reliable basis for the in-depth analysis of urban and rural planning and improve the persuasiveness of the decision making [13].

The rapid development of remote sensing technology and gradually maturing of Geographic Information System technology applications provide the foundation for urban-rural planning from the qualitative analysis to quantitative analysis. The core technology of the urban and rural planning is land use suitability comprehensive evaluation. The land use suitability assessment reflects the progress and consideration of the relevant natural, economic, social and technical data. It can be achieved by using high technology outcomes like Remote Sensing and Geographic Information System as well as other suitable existed methods for an individual assessment of every impact factor for protection and development. At the same time, it makes necessary to undertake a comprehensive and detailed analysis of the factor. It is necessary to select the most optimal allocation principles and finally derive the rational allocation program of various elements and resources.

The use of evaluation method can be clearly determined the space relationship of strict protection - appropriate protection - general use - optimal use - key development in different policy conditions, to provide a solid theoretical foundation and serviceable content for urban and rural planning. The use of remote sensing technology, Geographic Information System technology in land suitability evaluation is a new technology and method in urban and rural planning. This approach takes an application of remote sensing and Geographic Information System technology for study urban and rural planning in land suitability comprehensive evaluation [14]. It has been used the information through land use interpretation of 2006 TM image and the natural limiting factor, the factor of economic construction, ecologically sensitive factor, ecological protection factor which affect land suitability evaluation through Geographic Information System analysis. The natural limiting factor includes terrain factor and landscapes factor; the factor of economic construction includes town factor and traffic factor; the ecological protection factor includes basic farmland protection factor, nature reserve, scenic area factor, and drinking water source protection factor. These factors are acquired from Remote Sensing method and Geographic Information System technology analysis. The use of Remote Sensing image interpretation results, residential area was extracted to carry out. On the base of the characteristics of population distribution, using the natural impact factors (terrain) and socio-economic impact factors (towns scale, the distance from the town and the main network traffic and communication system), the population on the grid is discreet by spatial analysis functions of Geographic Information System. Through analysing the various factors, which impact on land suitability and the distribution of population the weight of the various factors, which impacts on the development of urban and rural, evaluates the land use suitability to determine the direction of urban and rural has been determined. It is the scientific approach of the planning process [15].

As an important fundamental work in urban and rural planning, land use, suitability assessment provides important influence for planning, planning management, planning implementation, and planning evaluation. It opens an opportunity in-depth study and explore at the approaches of land suitability assessment, especially, with the use of Remote Sensing methods and Geographic Information System technology. In general, the study focuses on the use of Remote Sensing method and Geographic Information System technology as evaluation tools in urban and rural planning.

2.5.1. Technology and Methodology Description

One of the main factors affecting urban land use suitability assessment includes the natural limiting factor (terrain factor and landscape factor), the factor of economic construction (town factor and traffic factor), ecologically sensitive factor (ecological factor) and ecological protection factor (basic farmland protection factor, nature reserve and scenic area factor and drinking water source protection factor) [16]. Firstly, using object-oriented information classification technology, the construction land, water, woodland, grassland, unused land, arable land and fired land, and other indicators are extracted from TM remote sensing image, and based on these indicators relevant factors can be accessed through based-on Geographic Information System factor analysis. Secondly, it is to extract terrain factor from DEM and other basic geographic information. Thirdly, it is to extract nature reserve and scenic area factor, drinking water source protection factor, and other factors from planning and basic geographic information. Finally, it is to incorporate the evaluation model to comprehensive foregoing various factors for land use suitability assessment.

2.5.2. Data Interpretation

Land use is the main basis for urban and rural planning; the distribution of various Land Use/Land Cover types provides considerable constraints to urban and rural planning. In the land use, suitability assessment can be applied an object-oriented information classification technology to the classification of TM images. The statistic behaviour of the land classification is illustrated in the Table 6 as a typical example.

2.5.3. Terrain Factor

As per mentioned the terrain factor is very important in the process of the classification. It determines that the altitude of the land Use/Land Cover for urban and rural development and construction facilities – industrial or social is important while the access of data available for further decision-making. Therefore, it is highly vital to have an access into the terrain data for being able to classify successfully in the future design and construction process.

Table 7 reflects an example of terrain factors for the selected area demonstrating the method of performance of the factor.

Table 6. Statistics of land use classification

Classification	Area (square kilometres)	Percentage of total area
Arable land	1486.9	78.2%
Woodland	304.9	16%
Grassland	36.7	1.9%
Construction land	21.2	1.2%
Development land	2.8	0.1%
Water	47.5	2.5%
Unused land	0.3	0.02%

Table 7. Terrain factor classification

Classification	Extent	Score
Unlimited region in terrain	Surface elevation between 196 and 300 meters	10
Weak limited region in terrain	Surface elevation between 300 and 350 meters	8
General limited region in terrain	Surface elevation between 350 and 400 meters	5
Strong limited region in terrain	Surface elevation between 400 and 500 meters	3
Stronger limited region in terrain	Surface elevation more than 500 meters or less than 196 meters	1

2.5.4. Landscape Factor

Surface slope is one of the most important elements for impacting urban and rural development and construction, which not only affects the spatial structure within the city layout, but also to the economic investment and engineering measures. The landscape factor's classification table in Table 8 [17] grades landscape.

Table 8. Landscape factor's classification table in land suitability assessment

Classification	Extent	Score
Unlimited region in Landscape	Surface scope less than 5 degrees	10
weak limited region in Landscape	Surface scope between 5 and 10 degrees	8
general limited region in Landscape	Surface scope between 10 and 15 degrees	5
strong limited region in Landscape	Surface scope between 15 and 25 degrees	3
stronger limited region in Landscape	Surface scope more than 25 degrees	1

2.5.5. Town and Traffic Factors

Towns are main space gathering regions of socio-economic elements, the location, size and grade of which play a decisive role in the distribution of the future population and the space layouts of urban and rural development.

A transport network has a strong guiding role in the urban and rural development, in accordance with the general experience of urban geography, towns, population and related elements of the productivity gathered into the open traffic elements (except railways and highways). Based on the results of Remote Sensing image interpretation, traffic factor is graded by the traffic factor classification [18].

2.5.6. Ecologically Sensitive Factor

Basic ecological protection factor is water system, wetlands, green forests and natural ecological protection zones and simultaneous nature reserves, scenic spots, key heritage conservation areas which are ecologically sensitive areas and must be taken seriously in urban and rural planning and construction.

2.5.7. Basic Farmland Protection Factor

According to the results of land use interpretation for the selected area for instance, for existing arable land area it has been classified the exceeding 1,500 square kilometres of arable land. Basic farmland as a part of arable land and as the best arable land must be strong protected. In the land suitability assessment taking into account the actual situation, the arable land with ground slopes less than 8 degrees and continuous area more than 50 hectares is taken as basic farmland [19-20].

2.6. Comprehensive Evaluation and Analysis of Results

Land suitability comprehensive assessment is to integrate various factors for affecting the various factors, add proper weight, and finally get comprehensive evaluation results of Land suitability. These factors include six flexible factors: terrain factor, landscape factor, town factor, traffic factor, ecologically sensitive factor, basic farmland factor, and two rigid factors: nature reserve and scenic area factor and drinking water source protection factor. The weight between flexible factors can be adjusted by local situation; rigid factor is to protect the local ecological environment and the safety of drinking water so that it is not randomly adjusted. The weight is mainly determined by expert knowledge and an experience accumulation in accordance with the importance of the indicators for an assignment.

Land suitability comprehensive assessment model is $Si = \Sigma ViWi$. Vi is the value of single-bound targets set in the formula. Its spatial distribution data can use space overlay data extraction, assign the value by experts based on Case analogue and Expertise, or operate the buffer zone and re-classification the single-bound.

The land suitability comprehensive assessment has been accessed basing on various factors of the different weight under the platform of ArcGIS, where they have to be calculated.

Considering comprehensive three rigid factors, which are nature reserve, scenic area and drinking water source protection, we can achieve the value of the land suitability comprehensive assessment for Liangping. It must be determined as appropriate construction sites if the value is more than six [21].

Remote Sensing is launched a successful application in urban and rural planning at all levels of Chongqing which is the strongest means of accessing the geographic information for its potential. Its technical superiority is superior to traditional methods for making the urban and rural planning. It provides a better platform for land suitability evaluation of the urban and rural planning which combines Remote Sensing and Geographic Information System.

This article made evaluation based on the weight analysis and we hope there may be further evaluation methods in future researches.

2.7. Remote Sensing Methods and Geographic Information System Technology Applications in Land Use Purposes for Engineering Solutions

Geotechnical works provide wide information related to the soil characteristics. One of the important soil's parameters is its electrical conductivity, which allows defining the amount of chemicals dissolved in the soil and groundwater. The values of electrical conductivity are important for designing the Earthing system and for defining the corrosive activity of the soil.

The study of salt condition of soil has a big importance for agriculture purposes. It relates for Land Use/Land Cover investigations where need to be conducted appropriate studies as the significant source for decision makers. Determination of the amount of salts dissolved in the soils is also important for construction needs. The type of concrete and protection for foundation depends on the amount of different chemicals dissolved in the soil and groundwater. Many parameters influence on the amount of chemicals in the soil. There is no any accurate data for Azerbaijan soils in common.

For the reason of Land Use/Land Cover investigation, the use of advance of the space technology has been suggested. The immediate result of the investigations is the inventory of land Cover/Land Use of the selected area of the Absheron peninsula and an output, which has an essential input to the land characteristics identification for the purpose of the agriculture planning.

The use and application of the space technology in a huge case in particular, for the case of Land Cover/Land Use studies is more suitable means due to the covering large areas, high accuracy, the availability of application in the unacceptability areas etc. Moreover, according to the created and developed database there is an advantage of being very sensitive to any available change occurred in the investigated areas. At the same time, the space technology creates an excellent environment to conduct and receive appropriate outcomes within a short time period, which is highly important issue in soil salt investigations.

It is obvious that integration data from the Remote Sensing method into the geographic information system based on conducted soil investigations with measured soil parameters provide comprehensive information for clarification of the soil condition. Based on processed satellite data with integration of filed studies there is an approach of investigation and identification of the salt degree monitoring divided into not less than three categories. It is an advantage of the development of Geographic Information System technology which plays a valuable place in Land Use/Land Cover problem solving as an excellent information source especially extremely important directory for the local authority decision makers.

2.7.1. Scope of Problem and Method of Measurement

As a part of geotechnical works electrical resistivity test is one of the easiest ways for the definition of the amount of chemicals dissolved in the soil and groundwater. The values of electrical conductivity are important for designing the Earthing system and defining the corrosive activity of the soil. Besides electrical resistivity test, chemical analysis of soil and groundwater also show the amount of chemicals which influence soil conditions and can be used for determining their corrosive activity.

By means of those investigations and parameters, characteristics of the soil for many of applications and engineering decisions can be done. For this reason, the study of soil is a very important issue for many purposes.

Electrical resistivity testing method is conducted by the use of two test electrodes with the resistances r_2 and r_3 and the third electrode with the resistance r_1. These electrodes should be located at some distance from each other so that test results can be trustworthy.

Figure 1. Site 1, Sangachal area.

Test consists of measuring the resistances between each pairs of electrodes. Final results show the values r_{12}, r_{13}, and r_{23}, where

$$r_{12} = r_1 + r_2 \text{ and etc.} \quad (1)$$

Value of r_1 may be found from results gained, so

$$r_1 = (r_{12}) - (r_{23}) + (r_{13})/2 \quad (2)$$

2.7.2. Chemical Analysis of Soil

The main parameter needed to be accepted from chemical analysis is the pH value which shows the level of corrosive activity of soils.

For determination of pH level in soil samples it is mixed with deionized water ratio sample and the pH of the resulting aqueous suspension is measured by pH meter with combination electrode and automatic temperature compensation.

Figure 2. Site 2, Garadagh district.

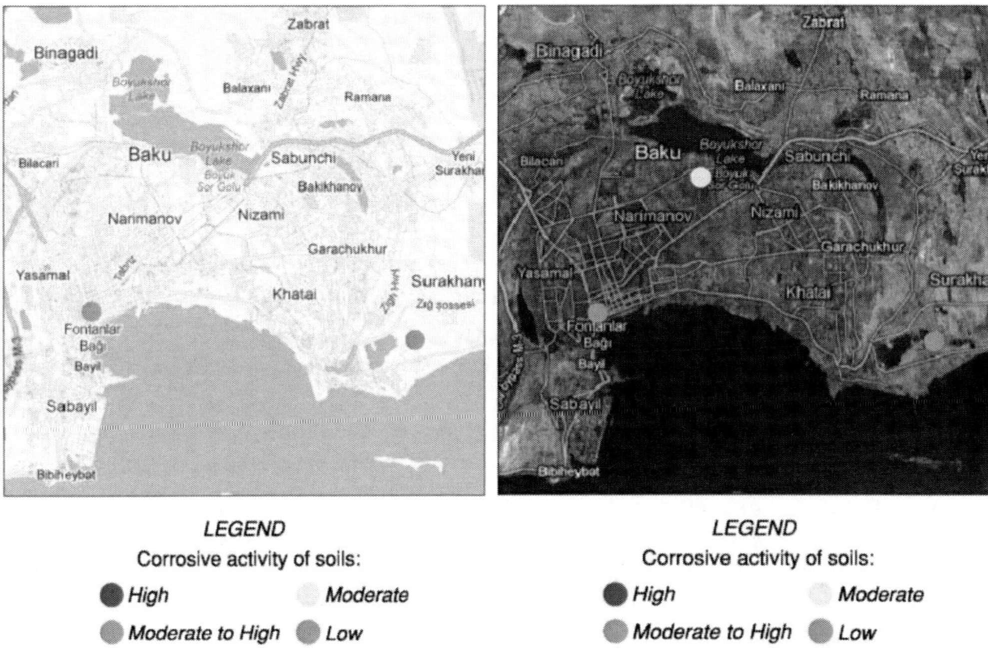

Figure 3. Site 3, Baku city.

In water samples, pH is measured by pH meter with combination electrode and automatic temperature compensation.

2.7.3. Outcomes

For studying the corrosive activity of Absheron Peninsula soils, the electrical resistivity test and chemical analysis results have been taken for one site of peninsula with further interpolation for other two sites where space technology methods were used. At the same time, the results accepted by the use of space technology were evidenced by the field measurements (see Figures 1-3).

Results of soil condition measurements are shown in the Table 9.

Table 9. Measurement outcomes of soil condition

Sites	Soil electrical resistance, Ohm*m	pH	Corrosive activity
Site 1	676.65		Low
	195-240	7.8-8.1	Moderate
		6.2-6.8	High
Site 2	15.8-76.9		Moderate to high
		7.0-8.4	Moderate to high
		7.1-8.9	Moderate to high
Site 3	240-464	8.6	Moderate to low
		7.2-9.7	Moderate to high
		6.5-8.8	Moderate to high

These are excellent results demonstrated successful use of the space technology advances for soil salinity investigations.

2.7.4. Methodology

Remote Sensing technology in combination with Geographic Information System can render reliable information on soil study. The analysis of the spatial extent and temporal change of soil using Remotely Sensed data is of critical importance for soil monitoring. Change detection has temporal effects as Variation in spectral response which involves the situations where the spectral characteristics of the soil and other land covers the type in the selected time [22].

Change detection and monitoring involve the use of multi-date images to evaluate differences in soil contamination due to environmental conditions and human actions between the acquisition dates of images.

For this reason, the soil study vegetation was identified as an indicator for a change detection. In conformity with this study various change detection techniques, spectral change analysis method with special emphasis on vegetation indices were selected. The Normalized Difference Vegetation Index (NDVI) was used for a vegetation change detection due to its acceptable accuracy and the ability to detect the green vegetation.

NDVI values, which extracted for space images from 2008 and 2012 years were compared statistically. The statistical analysis carried out on the NDVI values in different years showed that the reductions of NDVI values are different.

The application of indices NDVI series of images supported by the results of the classification were supervised and generated to thc Geographic Information System which indicates the vegetation degradation in the investigated area.

a. Study Area

The Garadagh is located between the southern outcrops of the Caucasus Mountain range and the Caspian Sea, some 35 km south of the capital of Azerbaijan Baku (Figure 4). Garadagh semi desert extends on 1080 km² with population 111 753 and is characterized by a semi-arid climate with continental influence and humid, cool winters and dry hot summers. The mean July temperature reaches about 26.0°C and the mean January temperature 2°C in this area. Average rainfall is 200-400mm per year but can be 150-200mm in semi-desert areas when the less rain comes.

b. Objectives

The objectives of this study include:

- Analysis of the Spatial-temporal change of vegetation in the selected area;
- Vegetation classification;
- Performance of NDVI calculation, showing vegetation reflectance in the selected area;
- Change detection of vegetation distribution from 2008 and 2012 in the selected area.

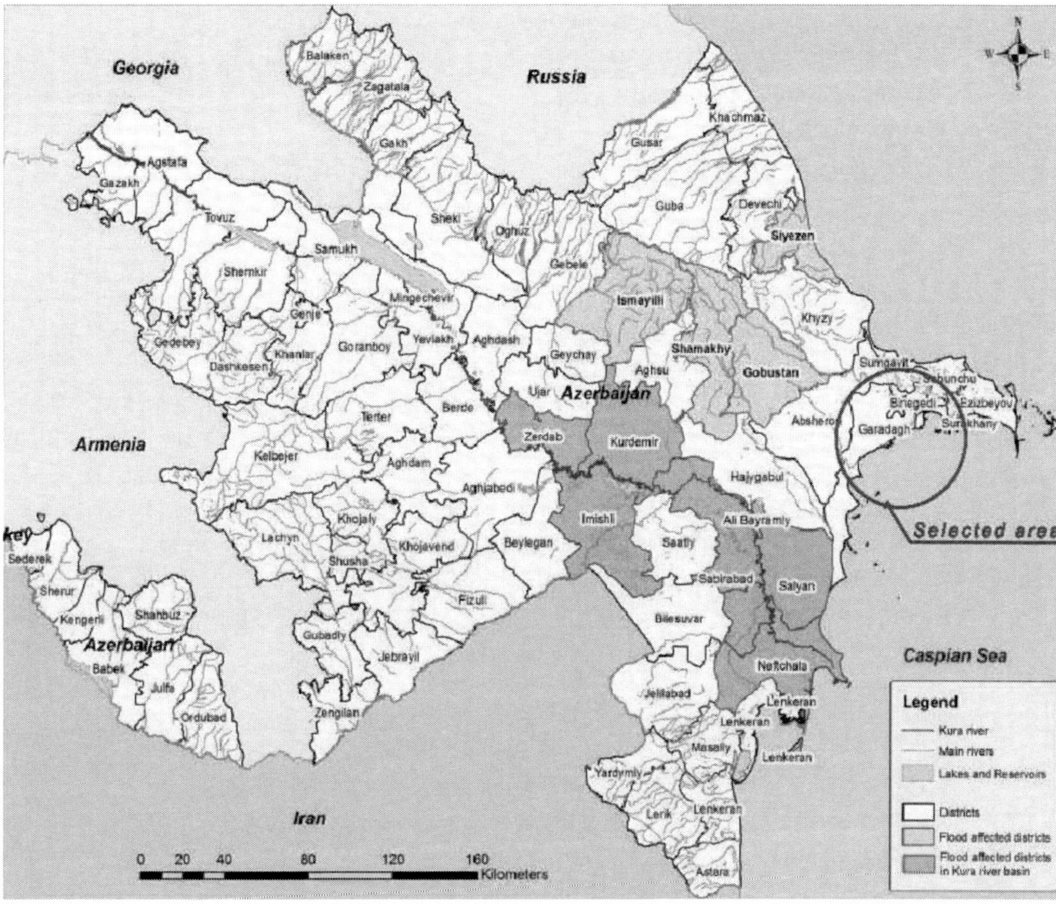

Figure 4. The map of selected area for study.

c. Methodology

High resolution image was used for processing between years of 2008 and 2012.

The methodological approach of this project includes training in the field of identification and census methods and the use of standardized monitoring methods. Remote Sensing (RS) and Geographic Information Systems together form a powerful information acquisition and analysis tool for monitoring environmental changes.

The steps of developments were following:

- Space image/classification and NDVI calculation;
- Classification/supervise of maximum likelihood;
- Classification map/NDVI map;
- Classification map/accuracy assessment of image;
- NDVI map/Vegetation change detection;
- Geographical data/Ground truth data-field data.

Analyses have been based on the main three methods of data processing:

- Maximum Likelihood classification
- NDVI Calculation
- Vegetation change detection.

From the supervised classification (Figure 4) of data, four classes can be identified and integrated into the soil salt degree classification.

Change detection and monitoring involve the use of multi-date images to evaluate differences in vegetation distribution due to environmental conditions and human impacts between the acquisition dates of images [23].

Change detection was taken into the changes which taken place on a period of 2008 to 2012. The Change detection is a thematic image, typically divided into the categories of Background, Decreased, Some Decreased, Unchanged, Some Increase, and Increased.

The study of soil salt condition is important for many purposes. Those studies needed to be provided within a short time of period for a large area. The space technology opens an opportunity to speed up the process of land condition as well as soil salt condition.

It has been defined the indicators for monitoring soil condition. Vegetation took indicator for selected area. A method ology for classification based on Remote Sensing with further GIS developments.

CONCLUSION

The method and methodology assignment depends on the task behaviour and features which are highly important in a success. In this regard, the demonstration of the selection of methods and methodologies has been undertaken paying attention to the segment and expecting to consider within the framework of data processing.

Therefore, this chapter reflects the methods and methodologies for the use of space science and technology advances in different areas of engineering solutions.

Chapter 3

SPACE IMAGE SOURCE FOR PROCESSING

ABSTRACT

This chapter undertakes the variety of sources of the data directed for an application in different areas of use. One of the valuable aspects of such sources is the Remote Sensing method. It considers the access of data from satellite or airborne keeping in appropriate form, which required in processing as the next stage of the process.

There is no doubt that data collection process has a continuous base. It has to be found out a certain way to access the required way of data satisfaction in the limited cases. This chapter demonstrates the capture of the information using multilevel data collection with application of naturally selected areas. It makes the integration of the received information into the space data as an evidence of filed measurements possible.

Keywords: space and airborne data, multilevel measurements, energy transmission, wavelength, data processing with field information integration

INTRODUCTION

There are several types of data based on Remote Sensing methods. Obviously, the use of electromagnetic wavelengths opens a wide application for a high accuracy of the data access. It is mainly suitable in space science and technology applications in the Earth studies. It can be used the Earth's own radiation or reflection of radiation from the Earth directed from the distance. Those sources reflected from the Earth are excellent data in further processing stage. This method and approach makes possible to study and classify the details of the Earth features and segments. At the same time, such approach creates wide applications depended on the expected task solving.

3.1. Remote Sensing Methods and Geographic Information System Technology Bases and Evolution Stages

As per it was indicated above Remote Sensing identified as a source of information obtaining from an object, an area or any features of the Earth surface without coming in direct contact with investigated subject. Based on this approach of Remote Sensing, a number of things would be coming under Remote Sensor, e.g., seismographs, fathometer etc. Without coming in direct contact with the focus of earthquake, seismograph can measure the intensity of earthquake. Likewise, without coming in contact with the ocean floor, fathometer can measure its depth. However, modern Remote Sensing means acquiring the information about the Earth's land and water surfaces by using reflected or emitted electromagnetic energy [24].

In conformity with the definitions from above point of view, the stages of Remote Sensing methods' use and applications can be described:

- the use of different types of sources of electromagnetic radiation (EMR) (Sun, artificially used EMR);
- transmission of energy from the source to the surface of the Earth, through atmosphere;
- interaction of EMR with the Earth's surface;
- transmission of energy from surface to Remote Sensor mounted on a platform (air or satellite based), through atmosphere;
- detection of energy by the sensor;
- transmission of sensor data to ground station;
- collection of data by ground station from mounted platforms (aero or satellite based);
- processing and analysis of the sensor data;
- final data output for various types of application;
- Final data output delivery to interested authority.

3.2. A Short History Description of Remote Sensing Methods

The knowledge about the history of Remote Sensing is necessary for better understanding of the subject, its scope and future development, particularly for the welfare of the human society. The development of Remote Sensing over time can be broadly divided into following six phases [1].

First Stage

Obviously, the human was always inquisitive about the things being reached or visible, for instance, deep in the forest or a mountain. It made human tension to try to reach greater heights, such as treetops, mountains etc. to observe phenomena on the Earth surface, to decide habitat places, farming and other day-to-day activities. The initial photographs of the Earth were captured from elevated platforms on the surface of the Earth. However, the actual beginning of Remote Sensing can be traced back in 1958, when free balloons were used for photography by the French Gaspard Felix Tournachon (known as Nadar) to photograph the village of Petil Becetre near Paris. In 1860, a part of Boston and Massachusetts were photographed from a captive balloon at 1200 feet height in USA. In 1909, the first aerial photograph was taken from an aeroplane, piloted by Wilbur Write over Centocelli, Italy. World War I provided a boost in the use of aerial photography. During that time, aerial photographs were used for two purposes – spying and mapping.

Second Stage: Development of Platforms and Sensors

Improvement in aeroplanes, cameras, films etc. resulted in the development of aerial photography during this period. The proper planning of flight for photographic purpose was also started. Topographical mapping was the main thrust upon the aerial photography. However, a number of scientists like geologists, botanists, soil scientists, geographers began interpreting the photographs to get information of their interest, especially for the development of natural resources. During this period the photographic coverage were increased both on the large and medium scale. World War II gave a real boost to photo interpretation technique, which was widely used for military intelligence purposes. The mapping of strategic location, military targets, and assessments of damages could be done accurately.

Third Stage: Development of Teaching and Training

After World War II, much emphasis on teaching and training of this technique was given due to previous experience of its wide use in different spheres. Many courses on Remote Sensing were started in reputed universities of the United States and Western European countries. International Geographical Union (IGU) set up a commission for the utilization of aerial photographs in 1949. The members of the commission emphasized the need of knowledge of those parts of the world, which were not before photographed and attention was given to cover more area by aerial photographs and techniques that were essential for interpretation.

Fourth Stage: Development of Instruments for Interpretation

In this phase, the techniques of photo interpretation became much more important in applying technique capability. A number of instruments was developed and introduced for interpretation during this period, which might be termed as a landmark in the progress of these techniques. It opened a new horizon for accurate and fast analysis and for monitoring the changes. Hence, a considerable advanced interpretation was made in many disciplines such as Geography, Geology, Geophysics, Agriculture, and Archaeology.

Fifth Stage: Significant Phase

This phase is very significant in the history of Remote Sensing as artificial satellites were launched in the space for acquiring information of the Earth surface. Though two American satellites, i.e., Explorer I and II were launched in 1958 and 1959 respectively under Explorer and Discover Programme, they were not important from Remote Sensing point of view. On 1st April, 1960, one satellite of eight members of TIROS (Television and Infrared Observation Satellites) family was launched as a research and development project. As TIROS's name suggested, the satellite carried two types of sensing devices – firstly, television, camera etc. which took picture of the visible spectrum; and secondly, infrared detectors which measured the non-visible part of spectrum and provided information of local and regional temperature of the Earth's surface. The supply of Remotely Sensed data on the Earth surface was greatly increased with the launching of ERTS-I (Earth Resources Technology Satellite) on 23rd July, 1972. It was placed in a sun-synchronous polar orbit, about 600 miles above the Earth surface. It made 14 revolutions in a day around the Earth and its sensors were covering a series 160km wide strip. Then it was followed by ERTS-2 in 1975. With the launch of this satellite, the name of these satellites was changed from ERTS-1, 2 to LANDSAT-1, 2 respectively. Four other satellites in these series were launched one after another in this phase, with improved cameras and sensors. Beside this, European and Asian Countries launched many other satellites in the space during this period.

Sixth Stage: Recent Development Phase

In this period, Remote Sensing technique has been improved in two ways. Firstly, there have been developments of sensors, which can use infrared and microwave spectrum other than visible spectrum to get information about the Earth's surface. Secondly, there have been very important advances with respect to the platforms in which sensors are mounted. Besides, satellites have been launched for specific purposes and with specific capability. The ground resolution has been continuously increasing until today. Hence, the interpretation and mapping are becoming very easy, accurate, and purposive. The European Radar satellite (ERS-I) launched in 1991 opened the avenue for systematic global observation in the microwave region. The French Satellite 'SPOT' is producing the imagery to provide the three dimensional view under stereoscope. The satellite – IKONOS, launched on September 24, 1999 has 1 m. resolution in panchromatic and 4 m. resolution in multi-spectral cameras. USA, France, and India have planned a series of satellites, with improved capability, so that the users are assured of continuity of the data.

3.3. Facilities of the Remote Sensing Data Access

3.3.1. Platforms

The base, on which Remote Sensors are placed for acquiring the information about the Earth's surface, is called platform. Platforms can be stationary like a tripod (for field observation) and stationary balloons or mobile like aircrafts and spacecraft. The types of platforms depend upon the needs as well as constraints of the observation mission. There are three main types of platforms, namely:

1. Ground borne;
2. Air borne; and
3. Space borne.

1. Ground borne platforms: These platforms are used on the surface of the Earth. Cherry arm configuration of Remote Sensing van and tripod are the two commonly used ground borne platforms. They have the capability of viewing the object from different angles and are mainly used for collecting the ground truth or for laboratory simulation studies.

2. Air borne Platforms: These platforms are placed within the atmosphere of the Earth and can be classified into balloons and aircrafts in future.

a. *Balloons*: Balloons as platforms are not very expensive like aircrafts. They have a great variety of shapes, sizes, and performance capabilities. The balloons have low acceleration, require no power, and exhibit low vibrations. There are three main types of balloon systems, viz. free balloons, Tethered balloons, and Powered Balloons. *Free balloons* can reach almost the top of the atmosphere; hence, they can provide a platform at intermediate altitude between aircraft and spacecraft. Free balloons can lift thousands of kilograms of scientific payloads. Unless a mobile launching system is developed, the flights can be carried out only from a fixed launching station. The free balloons depend on the meteorological conditions, particularly winds. The flight trajectory cannot be controlled. All these make extremely difficult to predict whether the balloons will fly over the specific area of interest or not.

b. *The Association for Geographical Studies*. The entities with the facility for balloons have been set up in some countries. In general, tethered balloons connected to the Earth station by means of wires having high tensional strength and high flexibility. The tethered line can carry the antenna, power lines and gas tubes etc., but when the wind velocity is less than 35 km. per hour at the altitude of 3000m., sphere type balloon is used [25]. When the wind velocity is less than 30 km per hour, natural shape balloons are restricted to be placed. Tethered balloons have the capability of keeping the equipment at a fixed position for a long time and thus, useful for many Remote Sensing programs. *Powered balloons* require some means of propulsion to maintain or achieve station over a designated geographic location. These can be remotely controlled and guided along with a path or fly above a given area within certain limitations [26-28].

3.3.2. Aircrafts

Aircrafts are commonly used as Remote Sensing for obtaining aerial photographs.

The advantages of using aircrafts as Remote Sensing platform are high resolution of data recorded, the possibility of carrying large payloads, the capability of imaging large area economically, accessibility of remote areas, convenience of selecting different scales, an adequate control at all-time etc. However, due to the limitations of operating altitudes and range, the aircraft finds its greatest applications in local or regional programme rather than measurements on global scale. Besides all these, aircrafts have been playing an important role in the development of space borne Remote Sensing techniques. Testing of sensors and various systems and sub systems involved in space borne Remote Sensing programme is always undertaken in a well-equipped aircraft.

3.3.3. Space Borne Platforms

Platforms in space, i.e., satellites are not affected by the Earth's atmosphere. These platforms move freely in their orbits around the Earth. The entire Earth or any part of the Earth can be covered at specified intervals. The coverage mainly depends on the orbit of the satellite. It is through these space borne platforms that we get enormous amount of Remote Sensing data and as a result, Remote Sensing has gained the international popularity. According to the orbital mode, there are two types of satellites – Geostationary or Earth synchronous and sun-synchronous.

a. *Geo-stationary Satellites:* Geostationary satellites are the satellites, which revolve around the Earth above the equator at the height of about 36,000 to 41,000 km., in the direction of the Earth's rotation. They make one revolution in 24 hours, synchronous with the Earth's rotation. As a result, it appears stationary with respect to the Earth. These platforms always cover a specific area and give continuous coverage over the same area day and night. Their coverage is limited to 70 N and 70 S latitudes and one satellite can view one-third globe. These are mainly used for communication and weather monitoring. Some of these satellites are INSAT, METSAT and ERS series.

b. *Sun-synchronous Satellites:* Sun-synchronous satellites are the satellites which revolved around the Earth in the north-south direction (pole to pole) at the height of about 300 to 1000 km. They pass over the places on the Earth having the same latitude twice in each orbit at the same local sun-time, hence, are called sun-synchronous satellites. Through these satellites, the entire globe is covered on a regular basis and gives repetitive coverage on a periodic basis. All the Remote Sensing resources satellites may be grouped in this category. Few of these satellites are LANDSAT, IRS, SPOT series and NOAA, SKYLAB, SPACE SHUTTLE etc.

3.4. Sensors of Electromagnetic Radiation Detection for Remote Sensing

Remote sensors are the instruments, which detect various objects on the Earth's surface by measuring electromagnetic energy reflected or emitted from them. The sensors are mounted on the platforms discussed above. Different sensors record different wavelengths bands of electromagnetic energy coming from the Earth's surface. As for example, an ordinary camera is the most familiar type of remote sensor, which uses visible portion of electromagnetic radiation. Classification of Sensors Remote sensors can be classified in different ways as follows:

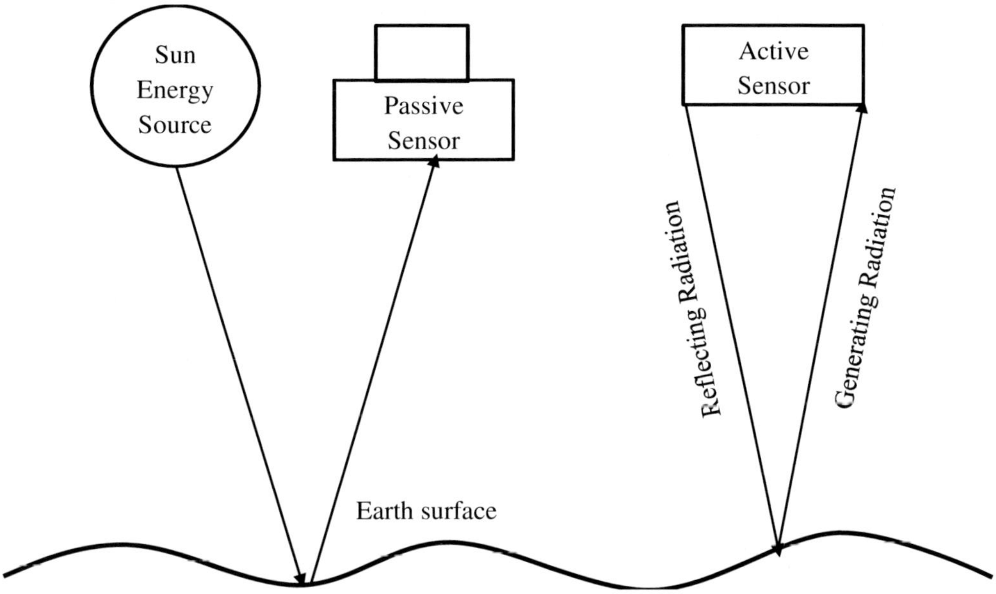

Figure 5. Type of sensors for Remote Sensing.

I. *On the Basis of Source of Energy Used:* On the basis of source of energy used by the sensors, they can be classified into two types:

- **Active sensors:** a. *Active Sensors* use their own source of energy and the Earth surface is illuminated by this energy. Then a part of this energy is reflected back which is received by the sensor to gather information about the Earth's surface (Figure 5).

When photographic camera uses its flash, it acts as an active sensor. Radar and laser altimeter are active sensors. Radar is composed of a transmitter and a receiver. The transmitter emits a wave, which strikes objects and is then reflected or echoed back to the receiver.

The properties of an active sensor are:

(i) It uses both transmitter and receiver units to produce imagery, hence, it requires high energy levels;
(ii) It mostly works in microwave regions of EMR spectrum, which can penetrate clouds and is not affected by rain;
(iii) It is an all-weather, day-night system and independent of the solar radiation;
(iv) The RADAR signal does not detect colour information or temperature information, but it can detect the roughness, slope and electrical conductivity of the objects under study.

- **Passive sensors:** b. *Passive Sensors* do not have their own source of energy. The Earth surface is illuminated by sun/solar energy. The sensor (Figure 5) receives the reflected solar energy from the Earth surface or the emitted electromagnetic energy by the Earth surface itself. Photographic camera is a passive sensor when it is used in the sun light, without using its flash. The properties of a passive sensor are:

(i) It is relatively simple both mechanically and electrically and it does not have a high power requirement;
(ii) The wavebands, where natural remittance or reflected levels are low, high detector sensitivities and wide radiation collection apertures are necessary to obtain a reasonable signal level. Therefore, most passive sensors are relatively wide band systems;
(iii) It depends upon good weather conditions.

II. *On the Basis of Function of Sensors:* On the basis of function of sensors, they are divided into two main types:

- **Framing System and Scanning System:**
  (i) Framing system: In framing system, two-dimensional images are formed at one single instant. Here, a lens is used to gather the light, which passes through various filters and then focuses on a flat photosensitive target. In ordinary camera, the target is a film emulsion, whereas in videocon camera, the target is an electrically charged plate.
  (ii) Scanning System: In scanning system, a single detector / a number of detectors with specific field of view, is used which sweeps across a scene in a series of parallel lines and collect the data for continuous cells to produce an image. Multi Spectral Scanner, Microwave Radiometer, Microwave Radar, Optical Scanners are few examples of the scanning system sensors.

III. *On the Basis of Technical Components of the System*: The sensors can be classified into three categories Based on the technical components of the system and the capability of the detection.

These are:

(i) Multispectral imaging sensor systems;
(ii) Thermal Remote Sensing Systems; and
(iii) Microwave radar-sensing systems.

The multispectral or multiband imaging systems may use conventional type cameras or may use a combination of both cameras and scanners for various bands of electromagnetic energy. As for example, Return Beam Videocon (RBV) sensor of Landsat uses both photographic and scanning systems, which is similar to an ordinary TV camera. The thermal system uses radiometers, photometers, spectrometers, thermometers to detect the temperature changes where microwave-sensing systems use the antenna arrays for collecting and detecting the energy from the terrain elements.

3.5. Remote Sensing Methods and Geographic Information System Technology Operation

3.5.1. Sources of the Electromagnetic Radiations in Remote Sensing and Geographic Information System

It is necessary to aware what kind of electromagnetic radiation (EMR) sources are used for data collection for the purpose of Remote Sensing. EMR is the dynamic form of radiated energy that propagates as wave motion equal to the velocity of light. The EMR is classified into different types based on their wavelength as follows in the Table 10.

Table 10. Electromagnetic Sources used in Remote Sensing

Kind of wavelength	Wavelength Range in micron
Cosmic Rays	< .0000001
Gamma Rays	.0000001 to .0001
X-rays	.001 to .01
Ultraviolet wavelength	.01 to .4
Visible wavelength	.4 to .7
Infra-red wavelength	.7 to 1000
Microwaves	1000 to 10^6
Radiowaves	more than 10^6

Out of these above types of EMR, Remote Sensing mostly uses the visible (0.4 to 0.7 microns), the reflected IR and thermal IR (.7 to 14 microns) and the microwave regions (1000 to 3000 microns).

There are mainly three sources of EMR used for Remote Sensing purpose, which illuminate the object. The Sun is the first and main source. The sun's radiation covers ultraviolet, visible, infrared and radio frequency regions. The maximum radiation occurs around .55 microns. All objects on the Earth surface also emit EMR. The wavelength of these EMR depends upon the temperature and other characteristics of the object. Many common objects emit EMR having wavelength between .85 and .95 um. It is also possible to produce EMR of a specific wavelength or band of wavelengths by the sensor to illuminate the object. This is called active Remote Sensing while the first two (EMR from the sun and self-emitted radiance by the object) are called passive Remote Sensing.

3.5.2. Transmission of Energy from the Source to the Surface of the Earth

While EMR is transmitted from the Sun to the surface of the Earth, it passes through the atmosphere. Here, electromagnetic radiation is scattered and absorbed by gases and dust particles. Besides the major atmospheric gaseous components like molecular nitrogen and oxygen, other constituents like water vapour, methane, hydrogen, helium, and nitrogen compounds play important role in modifying electromagnetic radiation. This affects image quality. The regions of the electromagnetic spectrum in which the atmosphere is transparent are called atmospheric windows. In other words, certain spectral regions of the electromagnetic radiation pass through the atmosphere without much attenuation are called atmospheric windows. The atmosphere is practically transparent in the visible region of the electromagnetic spectrum and therefore, many of the satellite based Remote Sensing sensors are designed to collect data in this region. Some of the commonly used atmospheric windows. They are:

(i) 0.38-0.72 microns (visible);
(i) 0.72-3.00 microns (near infra-red and middle infra-red); and
(ii) 8.00-14.00 microns (thermal infra-red).

The electromagnetic radiation when interact on the Earth, either is reflected, absorbed, radiated or transmitted through the material. It depends upon the nature of the object and the wavelength interacting with it. When the object is smooth in comparison with the wavelength of the incident radiation, it is reflected in the forward direction called specular reflection. When the surface is rough in comparison with the wavelength of the incident radiation, it is reflected uniformly in all directions, which is termed as diffused reflection. It may be noted that an object appearing rough in a particular wavelength region, is smooth in another wavelength region. As for example fine sand, which appears rough in the visible region, is smooth in the microwave region.

3.5.3. Transmission of Energy from Surface to Remote Sensor Mounted on the Platform

While EMR is transmitted from the surface of the Earth to the sensor, it again passes through the atmosphere. Here, electromagnetic radiation is modified again in the same way as it is transmitted from the Sun to the surface of the Earth (discussed earlier).

a. Detection of the Energy by the Sensor

The sensors like cameras or scanners mounted on platforms such as aircraft or satellite identify an object and discriminates it from other objects based on reflected/emitted electromagnetic energy received by it, which is called signature. The basic assumption is that each individual has a unique signature or fingerprint with which he can be identified. Likewise, every object has a unique signature on the sensor, by which it can be identified. In general, we can say that any set of observable characteristics, which directly or indirectly leads to the identification of an object, is termed as signature.

There are four major characteristics of objects, which facilitate discrimination:

- spectral;
- spatial;
- temporal; and
- Polarization variations.

Spectral variations are the variations in the reflectance or emittance of an object as a function of wavelength. Colour of objects is manifestation of the spectral variation in reflectance in the visible region. Vegetation looks green, because it reflects the 'green' part of the incident light preferentially. When the sensor uses wavelength region other than visible region, the objects do not give its natural colour as signature. As for example, vegetation looks red when the sensor uses near infrared reflectance.

Spatial variations in the reflectance and emittance are the attributes of the shape, size, and texture of objects. In the visible region of the spectrum, we are quite familiar with spatial information provided by the shape, size etc. of an object as seen in photograph. The photographs of a hot spring will give similar shape and proportion to the size of actual hot spring. However, the thermal scanner of a hot spring will give different look as per the temperature distribution across the surface of the hot spring.

Temporal variations relate to the reflectivity or emissivity with time e.g., the diurnal and seasonal variation. These variations play an important role in the detection of agricultural crops as well as soils. Two crops having similar spectral reflectance and having different growing season which can be distinguished by Remote Sensor. In case of rocks, diurnal variations in the emitted radiation (thermal infrared region) have proved to be very significant in their identification.

Polarization variation relates to the orientation of the electric field component of electromagnetic radiation. Objects can be distinguished from each other by making use of the differences in the polarization of the reflected EMR. Such observations have been particularly useful in microwave region.

3.5.4. Remote Sensing Data Processing and Output

When sensors like cameras or multispectral scanners are mounted on aircraft, the images are recorded on films or magnetic tape. When the aircraft lands, the films or tapes are physically transported to the laboratory for further processing. In case of satellites, there is no such possibility of physical retrieval of media containing data. In that case, the data is transmitted down to a ground station where it is recorded on magnetic tapes. The ground station therefore, needs to have an appropriate tracking antenna and communication link with the satellite [29]. Then the data is corrected in different methods to remove distortions caused due to motion of the platform relative to the Earth, platform attitude, the Earth curvature, non-uniformity of illumination, variations in sensor characteristics etc. This can be done either using electro-optical techniques or more popularly by using computer and is called processing of data. Then the output products are generated either in the form of photographs or in digital form. Then the photographs are analysed visually and digital data is analysed by computer software.

b. Data Products

The main interest of social scientists and applied scientists is the data produced by Remote Sensing technique. The Remote Sensing data are of two types – pictorial and digital. These data products are described in the following paragraphs:

Digital Data Products: The digital data products give information in the form of array of small cells having quantitative values, which is the function of the electromagnetic energy radiated from all objects within the field of view. A digital data product is called digital image. A digital image is a two dimensional array of pixels (picture elements). Each pixel represents an area on the Earth's surface and has an intensity value (represented by a digital number) and a location address (referenced by its row and column number). The intensity value represents the measured solar radiance in a given wavelength band reflected from the ground. The location address is a one-to-one correspondence between the column-row address of a pixel and the geographical coordinates (e.g., latitude and longitude) of the imaged location. The digital image is a vast matrix of numbers, is very often stored in a magnetic tape, and particularly in a computer compatible tapes (CCT). The digital data can be converted into photographic image.

Pictorial Data Products: The pictorial data products give information about the objects on the Earth surface in the form of photographs or images. The pictorial products provided by aircrafts are called aerial photographs. These are generally taken by sophisticated cameras, which use visible portion of electromagnetic energy. Therefore, aerial photographs give the exact view / picture of objects on the Earth surface on a reduced scale. The aerial photographs may be black and white or may be coloured; it depends upon the camera used in aircraft. The pictorial data products provided by satellites are called satellite images. These images are generally taken by sensors, which use both visible and invisible portion of electromagnetic energy. The satellite images can be black and white or coloured. The black and white pictures or images are produced from each band of digital data. For a particular band, black and white image is generated by assigning different shades of grey (white to black) to its digital data. Likewise, uncoloured images (blue, green, red etc.) can be generated by assigning different shades of blue / green/ red to a particular band data. When any three bands are combined, it gives multi coloured imagery. If images are taken in blue, green, and red bands (visible portion of electromagnetic energy) respectively, they can be combined to give a natural colour image. If images are taken in green, red (visible portion of electromagnetic energy) and infrared band (invisible portion of electromagnetic energy) and blue, green, red colours are assigned to them respectively, and then they are combined together, it will produce a False Colour Composite (FCC) image. The FCC image does not give the exact picture / view of the Earth's surface like aerial photographs. The layperson cannot visualize anything from FCC image. Only an expert can interpret it.

c. Data Interpretation

In the foregoing paragraphs, we have studied two major types of Remote Sensing data products, pictorial and digital. The pictorial data products, such as aerial photographs and satellite imageries are interpreted visually. Likewise, digital data products or digital images are interpreted mathematically by using computer software. Therefore, there are two ways of Remote Sensing data interpretation:

- Visual Interpretation and
- Digital Interpretation.

Visual Interpretation: Both aerial photographs and satellite imageries are interpreted visually. Photogrammetry is the science, which study interpretation of aerial photographs. To interpret aerial photographs, a number of sophisticated instruments such as pocket stereoscope, mirror stereoscope, plotter are used in photogrammetry for measuring area, height, slopes of different parts of the Earth photographed and for plotting different objects / themes from aerial photographs. With the development of science and technology, satellite imageries become more and more popular gradually. Satellite image interpretation is an art of examining images for the purpose of identifying objects and judging their significance. Interpreters study Remote Sensing image logically and attempt to identify, measure and evaluate the significance of natural and cultural features. Image interpretation technique requires extensive training and is labour-intensive. Information extraction from imageries is based on the characteristics of image features, such as size, shape, tone, texture, shadow, pattern, association etc. Though this approach is simple and straight forward, it has following shortcomings:

i) The range of gravy values product on a film or print is limited in comparison with what can be recorded in digital form;
ii) Human eye can recognize limited number of colour tones, so full advantage of radiometric resolution cannot be used;
iii) Visual interpretation poses serious limitation when we want to combine data from various sources.

Digital Interpretation: Digital interpretation facilitates quantitative analysis of digital data with the help of computers to extract information about the Earth surface [30]. Digital interpretation is popularly known as 'Image Processing.' Image processing deals:

- with image correction;
- image enhancement; and
- Information extraction.

Image correction means to correct the errors in digital image. Errors are resulted due to two reasons. When errors are resulted due to defect in sensor (as for example if one of the detector out of 'n' number of detectors does not work), it is called radiometric error. When errors are resulted due to the Earth rotation, spacecraft velocity, atmosphere attenuation etc., it is called geometric error. Both radiometric and geometric errors / noise in images are reduced through different techniques with the help of computer.

Image Enhancement deals with manipulation of data for improving its quality for interpretation. Sometimes digital image lacks adequate contrast, as a result of it, different objects cannot be recognized properly. Therefore, the image requires contrast improvement. Through different image enhancement techniques, contrast is improved in digital image. After image correction / rectification, and contrast enhancement, information is extracted from the digital image, which is the ultimate goal of an interpreter.

In *Information Extraction*, spectral values of pixels are analysed through computer to identify / classify objects on the Earth surface. In other words, spectrally homogenous pixels in the image are grouped together and differentiated from other groups. In this way, different features of the Earth are recognized and classified. The field knowledge and other sources of information also help in recognition and classification processes.

3.6. Data Sources for Remote Sensing and Geographic Information System Applications

It is obvious that appropriate data is required for an affirmative solution of problems and right decision-making. It can be achieved only through the collection of accurate, reliable and comprehensive set of data [31]. Remote Sensing methods have proved to be of great importance in acquiring data for effective resources management and hence, could be applied to environment monitoring and management [32]. Further, the application of Geographic Information System in analysing the trends and estimating the changes that have occurred in different themes helps in management decision-making process.

3.6.1. Remote Sensing Methods As an Instrument of Data Source

As it has been described previously Remote Sensing is the technique of deriving information about objects on the surface of the Earth without physically meeting them. This process involves making observations using sensors (cameras, scanners, radiometer, radar etc.) mounted on platforms (aircraft and satellites), which are at a considerable height from the Earth surface and recording the observations on a suitable medium (images on photographic films and videotapes or digital data on magnetic tapes).

Satellite Remote Sensing is widely used as a tool in many parts of the world for the management of the resources and activities within the continental shelf containing reefs, islands, mangroves, shoals and nutrient rich waters associated with major estuaries.

The mode can be geostationary, permitting continuous sensing of a portion of the Earth or sun-synchronous with polar orbit covering the entire Earth at the same equator crossing time. As an example, it can be so if the Landsat series of satellites have a repeat period ranging from 16 – 18 days. Satellites cover the same area and provide continuous coverage of a fixed area. The sun-synchronous satellites have a synoptic coverage. The ground area covered by the satellite's passes can be obtained by referring to its path and row.

Sensors are devices used for making observations. They consist of mechanisms usually sophisticated lenses with filter coatings to focus the area observed onto a plane in which detectors are placed. These detectors are sensitive to a particular region in which the sensor is designed to operate and produce outputs, which are either representative of the observed area as in the case of the camera or produce electrical signals proportionate to radiation intensity.

General parameters of satellite data can be reflected as following:

Spatial resolution: It is a measure of the smallest angular or linear separation between two objects that can be resolved by the sensor. The greater the sensor's resolution, the greater the data volume and the smaller the area covered. In fact, the area coverage and resolution are inter-dependent and these factors determine the scale of the imagery.

Spectral resolution: It refers to the dimension and number of specific wavelength intervals in the electromagnetic spectrum to which a sensor is sensitive. Narrow bandwidths in certain regions of the electromagnetic spectrum allow the discrimination of various features more easily.

Temporal resolution: It refers to how often a given sensor obtains imagery of a particular area. Ideally, the sensor obtains data repetitively to capture unique discriminating characteristics of the phenomena of interest.

Radiometric sensitivity: It is the capability of differentiation of the spectral reflectance/emittance from the various targets. This depends on the number of quantization levels within the spectral band. In other words, the number of bits of digital data in the spectral band will decide the sensitivity of the sensor.

3.6.2. Development of Geographical Information Systems

A Geographic Information System is a system of hardware, software, and procedures to facilitate the management, manipulation, analysis, modelling, representation, and display of geo-referenced data to solve complex problems regarding planning and management of resources. Functions of Geographic Information System include data entry, data display, data management, information retrieval, and analysis. The applications of Geographic Information System include mapping locations, quantities and densities, finding distances, mapping, and monitoring change.

Function of an Information system is to improve one's ability to make decisions. An Information system is a chain of operations starting from planning the observation and collection of data, to store and analysis of the data, to the use of the derived information in some decision making process. A Geographic Information System is an information system that is designed to work with data referenced to spatial or geographic coordinates. A Geographic Information System is both a database system with specific capabilities for spatially referenced data, as well as a set of operations for working with data. There are three basic types of Geographic Information System applications, which might also represent stages of development of a single Geographic Information System application.

a. Inventory Application

Many times, the first step in developing a Geographic Information System application is making an inventory of the features for a given geographic area. These features are represented in Geographic Information System as layers or themes of data. The emphasis at this stage of application development consists of updating and simple data retrieval.

b. Satellites and Sensors

Upon completion of the inventory stage, complex queries on multiple layers can be performed using spatial and spatial analysis techniques.

c. Management Application

More advanced spatial and modelling techniques are required to support the decisions of managers and policy makers. This involves shifting of emphasis from basic geographic data handling to manipulation, analysis and modelling in order to solve real world problems.

d. Uses of Coastal Geographic Information System Packages

1. Retrieval of information of any specific site.
2. Provision of summary data for planning purpose.
3. Graphical display for educational and public relation exercise.
4. Classification of coasts and management zone.
5. Predicting modelling to determine.
   - i. Projections of coastal changes.
   - ii. Impacts of individual schemes.
   - iii. Impacts from changes of use.
   - iv. Impacts from natural calamities.

e. Application of Remote Sensing and Geographic Information System in Coastal Ecosystem Management.

Based on Remote Sensing, a variety of data pertaining to the coastal zone like, identification of plant community, biomass estimation, shoreline changes, delineation of coastal landforms and tidal boundary, qualitative estimation of suspended sediment concentration, chlorophyll mapping, bathymetry of shallow waters, etc. can be collected and all these data will help in effective coastal ecosystem management.

The latest Indian satellites IRS – 1C, 1D, P4 and P6 with their improved spatial resolution (PAN – 5.8 m, LISS III – 23.6 m, LISS IV – 5.8 m, WiFS – 188 m and AWiFS – 56 m), extended spectral range (inclusion of middle infrared band in LISS – III) and increased repetivity (5 days for WiFS data) have opened up new applications in coastal zone.

Preliminary analysis of IRS – 1C, 1D data indicates that coral reef zonation, identification of tree and shrub mangroves, mudflats, beach, dune vegetation, saline areas, etc. as well as better understanding of suspended sediment patterns are now possible. The PAN data combined with the LISS – III and LISS - IV data are extremely useful in providing detailed spatial information about reclamation, construction activity and ecologically sensitive areas, which are vital for the coastal zone regulatory activities. The information available from merged PAN and LISS III, IV data about coral reef zonation, especially for atolls, patch reef and coral pinnacles, is valuable for coral reef conservation plans. The distinction between tree and shrub mangroves in FCC (middle infrared, infrared, and red bands) of LISS III provides vital information on biodiversity studies [33]. The high temporal resolution provided by the WiFS data is found to be a major improvement in studying the behaviour of suspended sediments in the coastal waters, which would help in understanding the movement of sediments and pollutants [34].

Satellite Remote Sensing has been found to be a very valuable application tool in forest management including mangroves, in not only monitoring, but also carrying out relevant observations, which can bring out the impact of deforestation on global climate. Remote Sensing of change detection is a process of determining and evaluating differences in a variety of surface phenomena over time. For detection of land cover change, multi temporal data of Landsat TM were found to be more suitable for identification of deforestation areas, mapping the regeneration/ regrowth of forest area and tracing major changes in land cover.

Remote Sensing data derived from different satellite such as Landsat MSS data can provide us with the information about the areal extent, conditions and boundary of coastal wetlands. Further, multi-dated satellite data can be used effectively to find out the changes in the aerial extent of mangroves. Sequential nature of IRS data provides opportunity to monitor changes in the land use activities in the mangroves. IRS has been quite extensively used for Mangrove land use pattern and both the visual and the digital analysis of IRS data provides useful information.

Data acquired by both passive and active Remote Sensing systems can be used in the study of mangroves. The research involves two main steps. In the first step, a broad classification of the general land cover, including mangroves is made (Figure 6). This indicates the best approach to dealing with the various types of imagery in order to detect the mangrove deforestation. The second step concentrates on the specific problem of detecting changes. It examines different approaches for monitoring the nature of the changes in order to produce maps showing the current and former conditions in an area. SAR data, which is independent of cloud, cover and weather interference can be used for mapping mangrove and estimation of mangrove biomass.

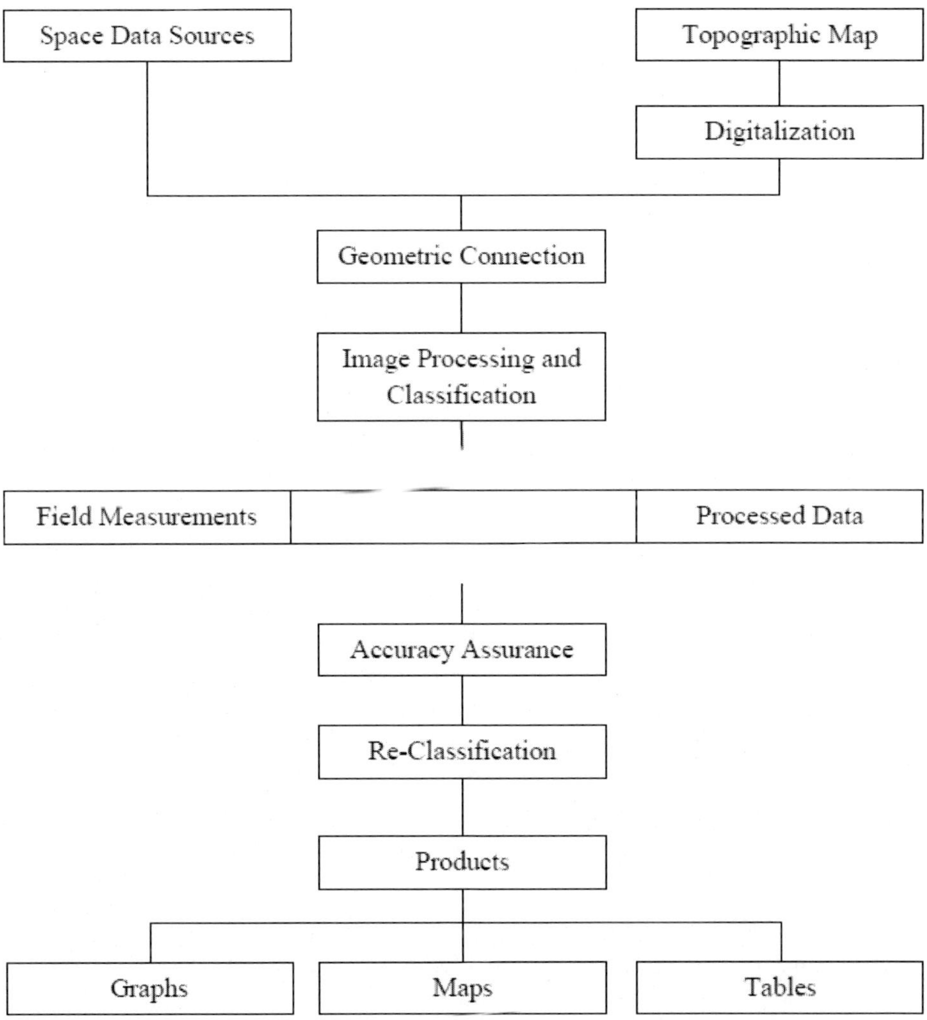

Figure 6. Flow chart for studying the ecosystem changes using satellite data [35].

One of the advantages of the use of Remote Sensing and Geographic Information System technology is mapping and change detection, for instance, identification of species/ plant communities, biomass estimation etc. A comprehensive database, including the information on distribution and extent of selected for investigation areas and species composition should be available for preparing a management plan. As a first step towards expected actions has to carry out mapping with reflection of dynamic changes of the features [36].

In addition, it could also be used to derive qualitative information about turbidity conditions in wetlands and studying turbidity is important for the health of the mangroves as it supplies nutrients to the system. As an example it can be referenced to the Tamil Nadu region in India where investigations were conducted for the area. In Tamil Nadu coastal zone, well-developed mangroves occur mainly in two places, one in Pichavaram and the other in Muthupet in the Cauvery delta. The mapping of wetlands in the Cauvery delta using Remote Sensing data showed that about 40% of the mangrove area has been degradedi.e., 32 sq. km. In 1976 has been reduced to 19 sq. km. in 1989 (IRS, 1991). However, the estimates using 1994 data showed that the area of mangroves is only 260 ha and nearly 63% has become barren between 1897 and 1994 [37].

Mangroves (India) and coral reef mapping along the coastal zone of Andaman Islands were attempted using SPOT and airborne SAR Xband imagery [38]. The availability of single polarization SAR data from aircraft and spacecraft platforms are not used effectively for mangrove mapping in India because of the difficulties in separating mangroves from other vegetation, influence of understory parameters, effect of soil moisture and the speckle effect.

The area covered by mangroves in the islands of Andaman was calculated using SPOT 1993 and IRS-1D LISS III 2003 imageries. The change in mangrove area within a span of ten years has presented in the form of a Table 11 (IOM report, 2003b) [39-41].

Table 11. Changes in Mangrove Area in Andaman Islands

S. No	Island	Area in ha (1993)	Area in ha (2003)	Change in area (ha)	% Change
1	Havelock	1468.51	895.37	-573.14	(-) 39
2	Neil	22.30	9.12	-13.18	(-) 59
3	North Passage	242.14	275.45	+33.31	(+) 14
4	Interview	1302.46	1116.26	-186.2	(-) 14
5	East	3.60	16.82	+13.22	(+) 367
6	Strait	18.28	19.33	+1.05	(+) 6
7	Ross & Smith	358.68	236.76	-121.92	(-) 34
8	Long	78.07	40.11	-37.96	(-) 49

3.6.3. Coral Reef Ecosystem

The coral reefs characterize an ecosystem of high biological diversity, having the greatest number of species of any marine ecosystem. Coral reefs are considered as one of the most important critical resources for various ecological, environmental and socio-economic reasons. They play an important role in global biochemical processes and in the reproduction of food resources in the tropical regions. They act as a barrier against wave action along coastal areas thus preventing coastal erosion. In addition, coral reefs protect mangroves and sea grass beds in certain areas, which are the breeding and nursing grounds of various economically important fauna. Because of the increasing human population along the coastal area, anthropogenic impacts on the coastal zone have become severe over the past few decades. Coral ecosystem also face many threats, of which some are of natural origin like storms and waves particularly tropical storms and cyclones that cause major intermittent damage to reefs. The majority of damage to coral reefs around the world has been through direct anthropogenic stress.

IRS LISS -II data has been used to map the coral reefs in the Gulf of Mannar on a much larger scale (up to 1:25000) under the DOD programme on Critical habitat Information system. Supervised classification of LISS -III digital data is used to identify the different species of coral reefs apart from the classification of odd patch reef, fringing reef and atolls. Due to high spectral resolution, (23m) LISS-III data were found to be more useful for coral reef mapping and monitoring. Digital image processing is widely used for coral reef mapping. IRS -LISS III data was processed using ERDAS IMAGINE software for coral reef classification and 50 classes were identified using the unsupervised cluster model of ERDAS. All the four bands of the image were used in the classification.

In Gulf of Mannar, the coral reefs have been used as a source of calcium carbonate and building blocks. In general, coral reefs in Gulf of Mannar can be categorized as "degrading," and hence, monitoring and management of these valuable marine resources are of prime importance. IRS LISS-II (1988), IRS LISS-III (1998) satellite data and ARC-INFO and ARC-VIEW GIS software have been used for coastal geomorphology, seafloor, shoreline, coastal land use/land cover and coral reef mapping of Gulf of Mannar.

The Remote Sensing data of SPOT, Landsat, IRS LISS -II and IRS LISS-III are used for the mapping of coral reef mapping. Using these data the coral reefs can be identified and mapped. The living and non-living corals appear in turquoise- blue and greenish-blue tone on SPOT FCC of band combination 2,3, & 4 [42]. In LISS -III the band combination 3, 2, & 1 are also found to be useful. At the time of validation of mapping, it is also observed that the tonal variations in coral reefs are not only due to the living and non-living corals, but also due to the depth of occurrence.

The coral reef area for the islands of Andaman was calculated using SPOT 1993 and IRS-1D LISS III 2003 imageries. The change in coral reef are within a span of ten years has presented in the form of a table [9]. Remote Sensing data (IRS IC LISS-III 1998) was used for mapping the coral reefs of the Kadmat Island in the Lakshdweep islands and digital classification was performed for the identification of corals around the island.

Very little information on maps showing the distribution of various benthic habitats in the coral reef is available. Remote Sensing technologies have a great potential in overcoming this information void.

Space Technology Application and Research Program (STAR) evaluated the suitability of latest satellite in Landsat series: Landsat 7 ETM+ for investigating coral reef habitat in several small islands in Riau Archipelagos in Indonesia.

3.6.4. Lagoonal Ecosystem

Coastal lagoon is defined as an inland water body, usually oriented parallel to the coast, separated from the ocean by a barrier, connected to the ocean by one or more restricted inlets and having depths which seldom exceed a couple of meters. Lagoons are formed due to the rise in the sea level during Holocene or Pleistocene period. They constitute a very fragile ecosystem in the coastal zone. They are important indicators of environmental degradation. The coastal lagoon systems exhibit very high primary and secondary production rates; provide nursery and breeding ground for many species of marine fauna. Recently the lagoons of these types are experiencing an accelerated decline in the water quality, an increase in siltation, environmental problems such as pollution of sediments and increased turbidity.

With the advent of satellite-based Remote Sensing platforms, the availability of geo-referenced information and the potential for use of such information have increased dramatically. There is a major information gap in the mapping and broad-scale investigation of the shallow-water coastal areas from which the majority of Pacific rural people derive their sustenance and income. Cartographers have concentrated on terrestrial areas (for land use planning) and on areas outside the lagoons (for navigation), and most maps and charts draw a complete blank over the maze of coral reefs where tropical fisheries are most accessible and most productive. This shallow-water area is also most amenable to investigation through the multispectral imaging capabilities of Remote Sensing satellites such as SPOT, where classification of digital images can yield extremely cost-effective, even priceless information to assist in the definition of living marine resource potential, and subsequent management.

Pulicat lake is the second largest brackish water lake in India and it is located between 13°26' and 13°43' N latitude and 80°03' and 80°18' E longitude lying almost parallel to the Bay of Bengal. It extends over the Ponneri and Gummidipundi taluk of Thiruvallur district in Tamil Nadu and Sulurpet and Tada taluk of Nellore district in Andhra Pradesh and covers an area of about 461 sq. km. A study was undertaken at the Institute for Ocean Management, Anna University to delineate the water spread area, lagoon configuration in conjunction with shoreline, water quality viz. salinity, suspended sediment concentration, saline water intrusion, bathymetry etc. of Pulicat Lake using IRS- LISS I, LISS II and LISS III data of 1989, 1990 and 1998. in this study all radiometric and atmospheric corrections were carried out using the VIPS32 functions in the pre-processing stage and a particular scan line of cloud free IRS-1A LISS-1 (non-monsoonal season) and Landsat MSS (monsoon) covering the entire stretch of Pulicat has been selected to study the spectral reflectance of the water body. Digital analysis of IRS and Landsat MSS data was carried out using VAX 11/780 image analysis system for finding out suitable enhancement techniques for various parameters of water quality [43-47].

False Fourier Transformation using IRS and MSS data enhances bathymetry (low and high depths) and these areas show good correlation with DN values of the raw data. Higher DN (reflectance) values in lower depths have also been contributed by the occurrence of sea grass beds, which have been verified by ground truth studies. Since IRS bands 4 and 2 have narrow spectral width when compared to MSS, the density-slicing ratio for the output of bands 4/2 provides qualitative information on suspended sediments concentrations. The colour variations derived from local optimization functions is expected to be due to chlorophyll concentrations. Resampling supervised classification based on field data provides information on salinity variations. The intrusion of seawater into the lagoon has been clearly shown using Principal Component analysis.

The Principle Component analysis of IRS LISS-I band 2 shows the lake configuration and the areas of submergence and erosion during monsoon (17-12-1990). At the end of the monsoon, the lagoon bar mouth will be completely exposed and it will be closed by the deposition of longshore drifted sediments during summer season. Hence, the mixing between the lake water and ocean is restricted, influencing increase in salinity, which will disturb the ecological health of the lake. Because of increase in salinity in the central part of the lake, the density of the water mass increases, which does not allow the suspended particles to settle at the bottom.

Coastal ecosystems are of great importance and of immense value to humankind in the present and in the future. They are being degraded at an alarming rate by various preventable activities including that of human interference. The coastal ecosystems are to be monitored periodically for better management plans. The satellite-based sensors provide valuable information useful in assessment, monitoring, and management of coastal ecosystems. Optical Remote Sensing data is very useful for mapping the coral reef, mangrove, and lagoon ecosystems. The information, which is thus derived, can be very useful in the coastal ecosystem management, which is greatly required for the sustainable use, development, and protection of the coastal and marine areas and resources. Thus, Remote Sensing and GIS technologies are widely used today in coastal ecosystem management.

3.7. Test Site/In-Situ Data for Earth Observation in Case of Limited Field Data Access

Remote Sensing is the method of making possible of observing, recording and collecting of objects or features on the Earth at far away distance places. In Remote Sensing methods, the sensors are not in direct contact with the Earth objects or features being observed. The information needs a physical carrier to travel from the objects or features to the sensors through an intervening medium. It makes available to receive output usually an image representing the scene being monitored. A further step of image analyses and interpretation is required in order to extract useful information from the image depending on the task nature needed to be executed.

The goal of a Remote Sensing method is to extract information that is directly related to a management process creating opportunity for efficiently adapting the data ultimately for interested purpose. It provides an impact in wide areas of time management as well as cost and financial resources coordination in the stages of development and implementation.

Remote Sensing is the advanced method of information collection in the large number of the Earth sciences and application studies. Data acquisition technologies and information extraction methods are pushed to give shift to user-oriented approaches where quantitative and reliable assessments, trend evaluations and forecasts are demanded. A success of this achievement depends on a successful use of multilevel data collected from space-borne and field observation systems. It is necessary to optimize the infrastructure for data acquisition and analysing, effectively data processing and integration with further use for decision-making. The option of optimization contains the adequate instrumentation, better timing and coordination in information collecting and management.

There is a limitation in some cases to the investigated area for filed data access. It makes necessary to embrace the need for a filed data using indirect way for satisfaction of space image processing needs. General approach is to use test site data integrated into the space data for processing.

In this paper it has been demonstrated Test Site/In situ data use for fitting of field data limitation in space image processing oriented for the Earth observation. One more advantage of offered approach is to use Test Site/In situ for multilevel data access. The use of country area with features containing the main land cover/land use needs has been recommended for filed data required in space image processing.

The Earth observation is used to monitor and assess the status and changes in the natural environment and to its protection if required. Currently the Earth observation has become technologically more and more sophisticated.

Accurate, timely, and reliable data are a precursor to analyse and study of several aspects of the integrated Earth system. Remote Sensing observations have catalysed several aspects of Earth system science by providing panoptic views and time-series measurements of the Earth study in ways nearly impossible to replicate by traditional ground-based methods. One of the primary goals of Remote Sensing method application is to study and understand all interacting and interfacing components of the Earth as a dynamic system influencing the Earth condition. The use of Remote Sensing method also become more important due to the significant influence of modern human civilization on the Earth planet and highly needs to minimize negative impacts along with the opportunities of the Earth observation, which provides and creates a success in study and improvement of the social and economic aspects of human life.

3.7.1. Multilevel Data Collection for Solving Data Lack Weaknesses

There is no doubt that space technology plays an important role in our daily life. A big number of Earth observing satellites are rotating in orbit. It demands to enhance observing capability of satellites. Up-to-date space science and technology advances make a possible accurate integration of different information incoming into one data which is highly important in implementation of the global observation systems targets. Remote Sensing information technologies have passed different stages within the development processes from sensor design to observation missions and data application. Approaches have been changed during these stages on the importance of the various links of the information chain. Today it is unambiguously recognized that the multilevel data collection is not only an integral part but also the backbone of a spatial data information system. In conjunction with the space component, it forms a common information structure, which creates affirmative approach to the data processing assurance. An adequately obtain and deploy information, substantial improvement is required in space technology application in multilevel surveys, observing systems and networks together with the development of global monitoring systems oriented information services.

It is necessary to calibrate the used instrumentations for validation including applied software systems (providing required quality data inputs) to increase and enhance the processed data accuracy through geo-reference and integrated ground-based field data [48-50]. One of the segments of data processing quality is data sharing and its integration from different levels (ground-based and Remote Sensing) sources, at different scales needed to increase the frame of the observations as well as making sure of cross-level and cross-thematic consistency of the acquired data [51].

The circumstances of data collection and processing are those high quality and timely delivery, successful integration and adequately adaptation into ground-based field data, limitation, and protection of definition during the development stage. The use of multilevel information system based on Remote Sensing data combined with ground-based field information is highly desirable. There is no doubt that any available information is the way for achievement of the best outcomes of expected production in data processing phase. It creates a very strong environment in high accuracy and reliability of processed database making easy the decision achievements.

How space technology can be effectively applied in multilevel implementations – space, test site and ground measurements? How it can be benefited when there is a lack of access into the test site or ground measurements?

An approach of using the indirect ways of data collection instead of test site or ground data when there are obstacles of access to the ground data for some reasons is the inherent part of space data processing.

3.7.2. Data Lack Weaknesses Problem Solving

The stages of space technology application demand required developments and implementations for final product. It has to be integrated the space image with ground-based data merged to the topographical map with appropriate map-scale which depends on the nature of the task expected to be solved.

There is importance of detailed elements and consequences needed to be undertaken within the frame of space technology data processing. One of the important issues of space data processing is data validation. In this regard, there are existing gaps limited of applications in the area:

- Obviously there is a big success in sensor technology but at the same time existing technological obstacles eliminate a possibility of their proper use;
- The lack of satisfied information for acquired data processing based on the field data with further Remotely Sensed data integration;
- Recognition and identification of space data and field data combination;
- Complication of integration of information from different sources;
- Interface problem between users and applications.

It is necessary to calibrate the used instrumentations for validation including applied software systems (providing required quality data inputs) to increase and enhance the processed data accuracy through geo-reference and integrated ground-based field data. One of the segments of data processing quality is data sharing and its integration from different levels (ground-based and Remote Sensing) sources, at different scales needed to increase the frame of the observations as well as make sure of cross-level and cross-thematic consistency of the acquired data [52].

The circumstances of data collection and processing are high quality and timely delivery, successful integration and adequately adaptation into ground-based field data, limitation, and protection of definition during the development stage. As it was indicated above the use of multilevel information system based on Remote Sensing data combined with ground-based field information is highly desirable. It creates a very strong environment in high accuracy and reliability of processed database making easy the decision achievements.

It is obvious that in some cases it is not possible to have appropriate data for processing for instance, ground-based, which limited space technology application in task implementations. It makes necessary to find out sources that can be embraced the lack of such information during the stage of space data processing.

3.7.3. Test Site/In-Situ Data Access Area Selection in Space Technology Applications

As the general way TS/In-situ is the artificially developed area, which allows the use of the area as ground-based field measurements with further purpose of integration into space data processing. It is required to use or develop if necessary such areas with the similar Land Use/Land Cover features of the investigated area in order to be able to merge ground-based filed data into satellite data. It is the demand of Remote Sensing methods applications in data processing. Obviously, the task of artificial TS/In-situ development is complicated and limited issue for a several reasons:

(i) to embrace main Land Cover/Land Use features of investigated area for classification;
(ii) required time for TS/In-situ development with the main Land Cover/Land Use classification reflecting investigated area;
(iii) the necessity of a huge number of artificial test sites development around the World.

How the problem can be solved with the comprehensive approach in order to eliminate obstacles when exists limited access to the ground-based field for data collection from the investigated area?

There is no doubt that it has to be developed or selected areas for the TS/In-situ use purposes with the large of Land Use/Land Cover features for classification which can be integrated into the space data as a main part of data processing.

Below we are describing and offering most suitable way of natural TS/In-situ use for this aim. It is important to emphasize that there are not a big number of papers dedicated to the test site studies and investigations.

3.7.4. Azerbaijan's' Natural Features for Test Site/In-Situ Data Purposes

Azerbaijan is situated on the Western Caspian coast, in the South Caucasus's East. The nine of eleven main climatic types of the Earth, from subtropics to alpine meadows, are represented in Azerbaijan. The four seasons can be watched simultaneously. There is a splendid flora of humid sub-tropics and everlasting snow and glaciers of Shahdagh and Murovdagh, Gabala forests, full of spring flavours, and hot steppes of Mil, Mugan, and Shirvan, and mist and rains over tea and citrus plantations in the foothills of the Talish Mountain Range and warm "Moryana" (a shore-to-sea breeze) and the Absheron Peninsula's gusty "Khazri" (the Caspian North wind), all neighbouring each other.

The Azerbaijan land is densely cut through by numerous, mainly highlands, rivers, most of which have noisy mountain-echoing waterfalls such as Afurji, Ghirintov and Sariguney, Mighi. Azerbaijan's nature represented by deserts and woods. Azerbaijan's flora is a rich and versatile.

Highland and plain lakes whose total number approximately equals 250 such as Tufan (the Caucasus Major), Alagellar, Gey-Gel, Maralgel, Zeligel, Karagel (the Caucasus Minor) truly adorn the Azerbaijan.

Azerbaijan's fauna is notable for the same richness and variety: more than 12,000 species populate the country. Goitered gazelles, wild boars, wolves, foxes, hares, coypu rats, pheasants, ducks, bald-coots ("kashkaldak"), and Sultan chickens inhabit the area. Deer, roe deer, wild boars, brown Caucasian bears, lynxes, wolves and, somewhere, leopards are typical for lowland and midland. The Bezoar, mouflon, ibex and others populate the highland, the birds being represented by pheasants, partridges, stone partridges, francolins, Caspian snow cocks, and Caucasian blackcocks.

The country appears to decline towards the Caspian Sea, where all rivers of Azerbaijan flow. Some of them enter the Kura, the country's main one, while others do into the Araks, which is the Kura's largest tributary. The largest lake in the world (about 400 square km of water surface) the Caspian Sea is unique in terms of its origins, natural and biological resources. The sturgeon, stellate sturgeon, beluga, salmon, kutum, shemaya, herring, lamprey, sheatfish, pike, chub and some others populate in the Caspian Sea and rivers inflowing deltas.

Climate: Borrowing rather small territory, Azerbaijan has unique natural-climatic features. From eleven existing climatic zones of nine are presented in territory of Azerbaijan: from subtropics up to the high-mountainous Alpine meadows. Mid-annual temperature is +13.7^{0}C, average temperature in January is +1.7^{0}C, average temperature in July is +27.9^{0}C. Climate is passing from temperate to subtropical. Strong northern winds mainly are common in autumn.

Relief: There are possible to allocate 4 basic parts in relief of Azerbaijan: the Big Caucasus (a part of the Main Caucasian Ridge), the Small Caucasian Ridge, Nakhichevan and the Kura-Araz lowlands, Talish Mountains and Lenkoran lowland.

The rivers and lakes: More than 1000 rivers are proceeding in the territory of the country. The largest river is Kura. Azerbaijan is famous for numerous sources of mineral waters. Most known are Istisu, Turshsu, and Badamli. There are no large fresh lakes in Azerbaijan, but many small lakes are present. Their number amounts to 250.

Flora: More than 4200 kinds of plants are counted in the territory of the republic. Some of them are unique and are typical only for Azerbaijan: Eldar's pine, a silk acacia, lignum vitae. Mountain woods are attractive for rest and tourism.

Fauna: The fauna of Azerbaijan is a rich and sundry. More than 12000 kinds of animals and birds live here. There are bears, turs, wild boars, goitered gazelle (jeyran), wolves, foxes are found here etc. The world of birds is also various in territory of the country. More than 250 kinds of birds live only in Kyzyl-Agaj reserve.

Existing nine climate zones from eleven, a rich relief, fauna and flora features make territory of Azerbaijan attractive for the Test Site/In-situ (TS/In-situ) area in data collection where limited ground-based field data access with a different reasons. The data accessed from TS/In-situ area can be used for identification and classification of different segments and features of the Earth surface which do not available directly from investigated area as an important element of reliability and confidence of space data processing.

The lack of the ground-based field data can be embraced using territory of country dividing into six TS/In-situ regions making available the use of data measuring and collecting appropriate Remotely Sensed data for further integration into space data processing stage. The exceptions of the Azerbaijan territory consists of a large of the Earth surface classifications thank to existing natural features. Figure 7 shows TS/In-situ areas with a rich content of the Land Cover/Land Use features.

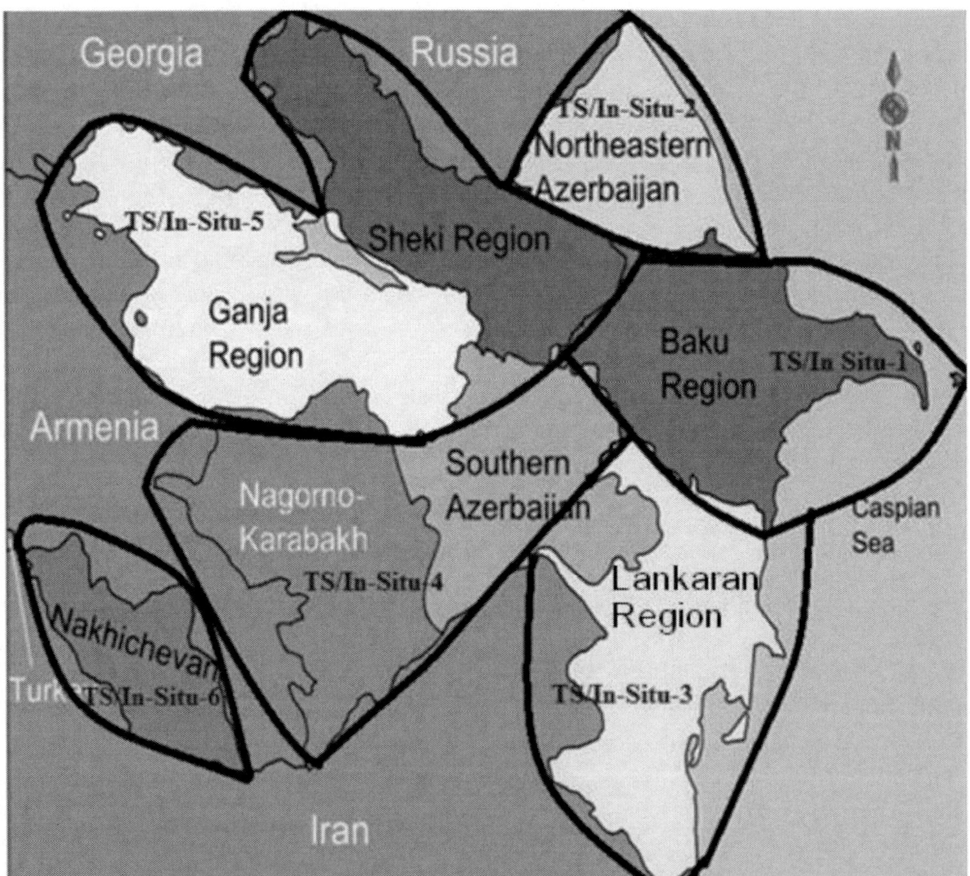

Figure 7. TS/In-situ data areas in Azerbaijan.

The Earth features in Azerbaijan territory divided into six country regions can be described as following:

TS/In-situ 1 (Baku region) – The Absheron Peninsula is located 29 m below World Ocean level. The climate of Baku and Absheron is known with warm semi deserts and steppes with Absheron-specific winds. Oil, gas, building stone, salt, sand are the mineral deposits of Absheron Peninsula. Salt lakes at the Absheron include Masazir, Gala, Beyuk-Shor and Hodzhasan. There are some of the oldest oil wells in the world where people would draw oil from with buckets and the first industrial oil wells on the Absheron.

Azerbaijan is the first among the world's countries by quantity and diversity of mud volcanoes. Among 800 mud volcanoes known in different countries on the Earth 400 are located within the boundaries of the South Caucasus oil-and-gas basins and among the latter 300 are located on the land of Azerbaijan, within its Caspian area of water and on numerous islands.

All known types of mud volcanoes on the world are represented in Azerbaijan. This is a genuine natural stock and laboratory of mud volcanoes. Due of this, in the last years Baku, the capital of the country has become a centre of international scientific forums on mud volcanism, geodynamics, and seismicity.

TS/In-situ 2 (North-eastern Azerbaijan) – it is the closest mountainous area to Baku, the capital of the country, among all the other mountainous areas of Azerbaijan. A visit to this region will give you the opportunity to see reach of types of forests with number of kind trees, famous Chirag-gala fortress and Khinalig settlement, home to indigenous tribes. In one of the districts, you will discover a cascade of waterfalls on the Gudialchay River in Tangaalti Gorge [53].

TS/In-situ 3 (Lankaran region) – The region is known as a sub-tropical area with extremely picturesque scenery and rich flora and fauna. For good reason, the Hirkan National Park and Gizilagaj Reserve have been created here. The gulf at the mouth of the Kura River, surrounded by cane thickets, attracts migratory birds, transiting in spring and autumn.

TS/In-situ 4 (Nagorno-Karabakh/Southern Azerbaijan region) – Excursions to the mountains and valleys with babbling brooks and granite blocks of varying and unusual shapes, which are scattered everywhere like the toys of a giant. Every opportunity for active vacations is presented in this region: from casual walks to extreme ascents of mountains and climbing frozen waterfalls... Thanks to its lakes and mountains the region is considered as one of the most naturally reach places in Azerbaijan. There are mighty forests, snowy peaks, noisy waterfalls, medicinal mineral spas, crystal-clear water, sprawling sub-alpine and alpine meadows.

TS/In-situ 5 (Nakhchevan region) – While traveling along the mountain ridge in the Nakhchievan Republic, we admire the stunning nature and historical monuments; and even drop in on settlements hidden in the valleys below. The cities of this land differ from the rest of Azerbaijan in their historical and artistic values, cosy side streets, and squares.

3.7.5. Method of Measurement

It is important to understand how to use and what method is required to apply for TS/In-situ data instead of ground-based information in space image processing. For full stages of information processing product based on space technology application the implementation of data acquisition and data configuring from the integration of remotely systems and ground-based field data is required. It is available or is to use TS/In-situ data when ground-based field data isn’t accessible.

A content of TS/In-situ systems for data acquisition is as following:

(i) geo-referencing for data merging and data calibration and validation;
(ii) spectrometric and other sensors for data acquisition;
(iii) Remote Sensing data gathering platforms development; and
(iv) Ground-based field data gathering facilities.

Azerbaijan territory can be used as TS/In-situ area instead of ground-based field data source when such a data is not available. The scale of application of the territory would be successfully covered within the worldwide framework. It is not necessary the spending of financial, technical, technological, or other required expenses for the purpose of ground-based field data access developing the number of TS/In-situ areas around the World. The territory of Azerbaijan can be commonly used and satisfied with technical and technological needs in ground-based field data for any countries involved on the Earth studies with the use of space technology or other applications.

1. Vehicle with equipment.
2. Equipment for data collection.
3. Platform for data collection.
4. Cable connecting data from monitored TS/In-situ to the data collection system.
5. Optical and receiving system for data collection from TS/In-situ.

Figure 8. Data collection from TS/In-situ area.

Figure 8 reflects stages and method of data collection starting from monitored TS/In-situ area up to input into the system as an initial phase of information collection with further integration into the space data. It is based on TS/In-situ receiving, collecting, primary mobile monitoring and processing system (RCPMMPS).

Description of the method: A mobile operation system can be used in each of selected TS/In-situ for the Earth monitoring purpose and it depends on the features of the area intended to be investigated. The data of TS/In-situ contains RCPMMPS and ground-based information which should be almost the same nature, as the studied area and for some reasons cannot be reached for ground-based data collection. The selection of appropriate TS/In-situ area from appropriate Azerbaijan regions gives an opportunity to achieve multilevel data collection – space image, RCPMMPS and ground-based. It makes available to complete and properly cover integration and processing stages of space technology application in the Earth observation and monitoring purposes.

An accuracy of the space data processing consists of the stages of data collection comes from successful merging and suiting of space information and data taken directly from investigated area. Honestly, the use of Remote Sensing method demands the confidence of space image processed data with ground-based measurements.

This paper demonstrates the use of RCPMMPS from TS/In situ instead of the data when access to the ground-based information limited for some reasons for instance, geographically or some other reasons of inaccessibility of area.

The territory of the Azerbaijan Republic divided into the six areas regionally can be mainly covered by the lack of the ground-based data and offered RCPMMPS which opens a wide opportunity in application of multilevel Earth observation and monitoring system in space data processing technologies.

CONCLUSION

Space technology application has a large place in our nowadays life. It is very difficult to limit the participation of space technology applications in the Earth studies.

This chapter describes the available data sources for space technology advances. There is a number of applications for using and accessing the data technology, which depends on the needs and requirements. It shows what needed to be done and selected to achieve expected targets.

Chapter 4

SPACE TECHNOLOGY APPLICATIONS IN ENGINEERING AND MANAGEMENT

ABSTRACT

The fact is that in any stages of the project execution the management has an important place. It relates from the point of tender up to as-build stages. For this reason, it is necessary to develop suitable system for data collection for further flexible use in any process of the project execution.

There is no doubt that information has a vital place in any management process. A Chapter IV reflects the stages of access and evaluation processes of information in management. It relates mainly engineering management areas. In particular, a comprehensive approach of disciplines and segments containing engineering activities has been undertaken.

The use of Remote Sensing with further Geographic Information System development makes possible a suitable application of achieved outcomes from the initial stage of engineering execution up to the end of project implementation. At the same time, it is very convenient to store the information and access at any time especially, in study of dynamic changes of constructed areas.

Keywords: Remote Sensing, Geographic Information System, engineering, engineering management, construction, engineering package

INTRODUCTION

The Construction Industry that has a huge number of tasks and costs involved in the projects is very large. The Project Managers have a hard time in monitoring of the projects between site and office. They have to come on site to know the progress of work and decide the sequence of work. They are generally confused on what to do next or what would be the changes faced by them in future. Therefore, the cost involved is large and it varies w.r.t. to the completion of the projecti.e., time. The project managers for planning who has a serious disadvantage for the decision making purposes as the spatial aspects fail to provide the required information which are still using the traditional approach for scheduling and progress control techniques such as bar charts and the critical path method. There is a pressure on the project managers to shorten the delivery times and thus the current scheduling and progress-reporting practices are in need of substantial improvements in quality and efficiency. Integration of Geographical Information System and project management software's with visualization was recognized as one of the most important tools for achieving this goal [54].

4.1. Remote Sensing and Geographic Information System in Management

An integration of Geographical Information System and Project management might assist a planner in better perception of a project as well as in the integration of other parties' activities in the planning process. Furthermore, in large scale projects, a visual representation of the schedule can be extended to monitoring not only the construction process itself, but also all the auxiliary activities, including onsite plant and equipment.

In addition, the practical and educational benefits of being able to visualize construction at a fine level of detail are significant. The application of geographic information system in project management will be new in the Indian Construction industry. A Geographical Information System will allow construction managers and different people involved in project with different backgrounds to get the information about the progress of the project and support Decision Making. A Geographical Information System will provide a common basis of understanding and communication among these people.

Many people think about Geographical Information System as a presentation tool. A Geographical Information System in fact, creates high quality maps that communicate considerable amounts of information in an efficient and attention-getting manner. A Geographical Information System is both a database system with specific capabilities for spatially referenced data, as well as a set of operations for working with the data [55-57].

Visualizing construction progress in three dimensions provides the construction project manager with a more intuitive view of the construction project manager and construction sequence. 3-D visualization allows the construction manager to view the construction activities during any stage of the construction process. The objective of this paper is to demonstrate that Geographical Information System can be integrated with project management software for construction progress visualization and an integrated information system.

4.1.1. Processes of Project Management and Their Interface

The term "Project Management" is defined as "The application of knowledge, skills, tools, and techniques for project activities to meet the project requirements." "The planning, monitoring and controlling of all aspects of the project and the motivation of all those involved in it to achieve the project objectives on time and to the specified cost, quality and performance." (PRINCE2) Project management processes can be organized into groups of one or more processes as shown in Figure 9.

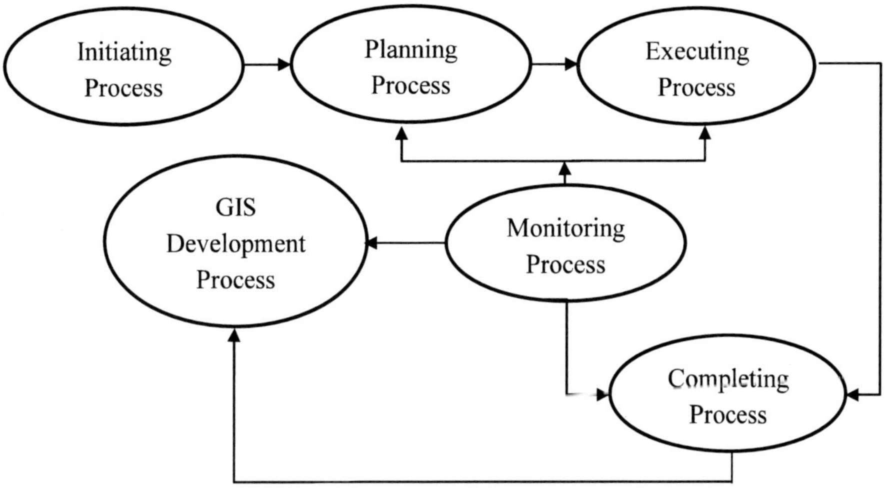

Figure 9. Process interface and applications between processes.

In conformity with indicated approach, it can be described as following:

- Initiating processes**:** recognizing that a project or phase should begin and committing to do so;
- Planning processes: devising and maintaining a workable scheme to accomplish the business need that the project was undertaken to address;
- Executing processes: coordinating people and other resources to carry out the plan;
- Monitoring processes: ensuring that project objectives are met by monitoring and measuring progress and taking corrective action when necessary;
- Completion processes: formalizing acceptance of the project or phase and ending it.

There is possible any other process needed to be undertaken for project management. It is the main elements of project management process required for consideration [58-61].

4.1.2. Geographical Information System Application in Management

In the past twenty-five years, a host of professions has been in the process of developing automated tools for efficient storage, analysis, and presentation of geographic data. These efforts have apparently been the result of increasing demands by users for the data and information of a spatial nature. This rapidly evolving technology has come to be known as "Geographic Information Systems." Geographic Information System goes beyond description; it also includes analysis, modelling, and prediction. According to the Environmental Systems Research Institute (ESRI), a Geographical Information System is defined as "an organized collection of computer hardware, application software, geographic data, and personnel designed to efficiently capture, store, update, manipulate, analyse, and display all forms of geographic referenced information." Kang Tsung Chang describes Geographical Information System as a computer system for capturing, storing, querying, analysing and displaying geographically referenced data. A Geographical Information System is essentially a marriage between computerized mapping and database management systems. Thus, a Geographical Information System is both a database system with specific capabilities for spatially referenced data, as well as a set of operations for working with the data.

Geographically referenced data separates Geographical Information System from other information systems. Let us take an example of road. To describe a road, we refer to its location (i.e. where it is) and its characteristics (length, name, speed limit etc.). The location, also called geometry or shape, represents spatial data, whereas characteristics are attribute data. Thus, a geographically referenced data has two components: spatial data and attribute data.

Spatial Data: Describes the location of spatial features, which may be discrete or continuous. Discrete features are individually distinguishable features that do not exist between observations. Discrete features include points (wells), lines (roads), and areas (land-use types). Continuous features are features that exist spatially between observations (elevation and precipitation). A Geographical Information System represents these spatial features on a plane surface. This transformation involves two main issues:

- the spatial reference system and
- the data model.

Attribute Data: Describes characteristics of spatial features. For raster data, each cell value should correspond to the attribute of the spatial feature at that location. A cell is tightly bound to a cell value. For raster data, the amount of attribute data is associated with a spatial feature that can vary significantly. The coordinate location of a Land parcel would be spatial data, while its characteristics, e.g., area, owner name, vacant/ built-up, land use etc. would be attribute data.

4.1.3. Geographical Information System Applications in Construction Industry

Geographical Information System applications have proliferated in the construction industry in recent years. This fact is illustrated by the growing number of articles finding their way into civil engineering and construction journals and conference proceedings, in addition to the handful of special publications devoted to Geographical Information System [62-63].

A Geographical Information System can be used for:

- Progress monitoring system in construction
- Networking solutions
- 3-D data analysis
- Site location and Client Distance
- Comparison of data
- Construction scheduling and progress control with 3-D visualization
- Government Regulations

4.1.4. Integrating Project Management and Geographic Information System

The intent of this project was to demonstrate the benefits of using Geographical Information System with the construction project management. In this project, the integration of Geographical Information System and Project Management is developed by using ArcGIS, MS Project, AutoCAD, and Visual Studio for assisting construction managers in controlling and monitoring construction progress. Successful project control is a challenging responsibility for all construction managers [64-66].

The objective of this project is to display the progress and sequence of construction work in 3-D while synchronizing this information with a formal CPM work schedule. This would help all parties involved in a construction project to visualize the progress in a natural way, hence, minimizing delays, and cost overruns. In addition to monitoring the schedule, the system can also be extended to monitor quantities of materials, costs, and resources.

Figure 10 shows the path of the project among the various applications in the system. It also shows the procedure that needs to be used in using the system. A building of G+ 20-store building is selected as the study area. The progress reports are described in the following sections.

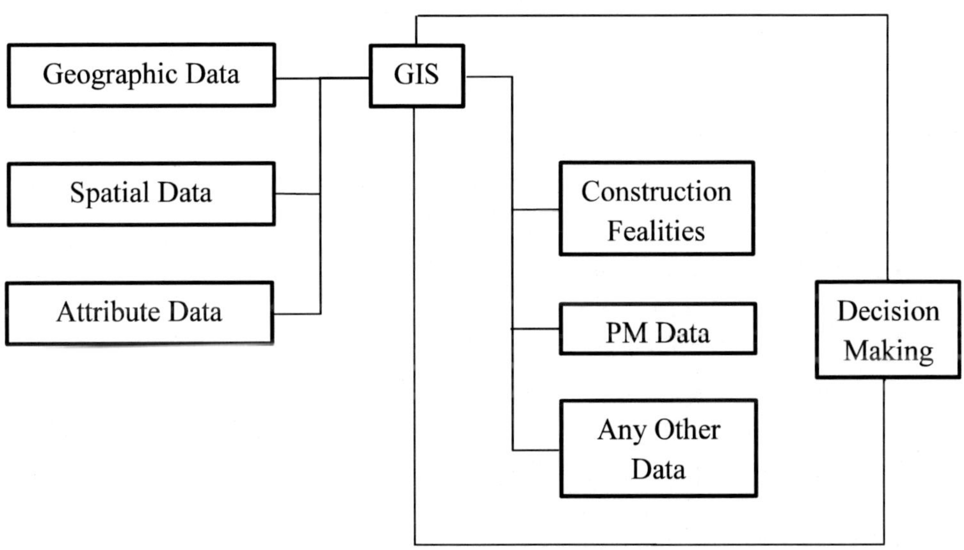

Figure 10. Steps of integration of project management data into Geographical Information System

4.1.5. Consistency of the Integration Process

- *Step I: Creating AutoCAD drawings – stage of construction features*
- *Step II: Creating work breakdown structure – stage of project management*

Activities in construction should be Mobilization of Site, which generally includes site clearance and preparing an access road for the site etc. The next activity would have to be RCC Work in which SubStructure. SuperStructure should be included into SubStructure which would be included as a part of construction of foundation. As the type of foundation for this structure is pile foundation, in which Piling has to be done where divided into different groups (GRP) [67-72].

According to the groups the numbers of piles are decided after the piles have been laid. The Pile caps have to be put on these piles simultaneously. On these pile caps a PCC Slab is to be laid which would make the pile foundation complete, i.e., the substructure activities would be completed. After the completion of substructure activities, superstructure activities have to be started. The type of formwork used is Mivan formworki.e., the concreting work is done at the same time for all columns, walls, beams, and slab.

Superstructure is divided into different floors, and then it is further divided into 2 typesi.e., wings (Type B, Type A). The two wings are then further divided into two partsi.e., in total 4 parts (Part I, Part II, Part III, and Part IV). Each part is again divided into structural partsi.e., BLDGColumn, Wall, Beam, and Slab. It is then planned that Shuttering and Reinforcement will be done simultaneously for BLDGColumn and Wall and then after that Staging for Beam and Slab will be done. Only after the whole of Shuttering and Reinforcement is done one can start the Concreting activity. In addition, after concreting is over one has to do De-shuttering of the formwork.

A list of activities in the construction stage is presented in Table 12. Each Activity for Slab, Wall, Beam, and Column are divided into components and they are further divided into individual activities.

Table 12. Information related to the construction (sample table)

Tasks / Document description	Project Name	Mhrs to be planned	Mhrs planned	Personnel		
				LR	NaB	TA
Reissue Steel Supports (GA, DTL, MTO)	Intelligent Nozzle Full Implementation	4	0			
Occupied area roof structural details with MTO	ST0094 Fire station	24	16			16
Slab (including inspection pit and drainage) GA/DTL/MTO	Fire Station Project	24	24	24		
Maintenance Platform Details and MTO	ACG FGC's Crossover Phase II	40	0			
Maintenance Platform Foundation Layout	ACG FGC's Crossover Phase II	20	8		8	
Maintenance Platform Foundation Calculation	ACG FGC's Crossover Phase II	16	16		16	
Maintenance platform foundation General Arrangement, DTL and MTO	ACG FGC's Crossover Phase II	40	16		16	
Structural Calculation	ACG FGC's Crossover Phase II	8	0			
Site Survey	Pig Wash	20	0			
Drawing change	Repair of subsided concreted area at SPS	4	0			

- *Step III: Initiating scheduling process*

MS Project 2003 was used as the scheduling tool. The project was scheduled based on the activities identified in the WBS described in Step 2. MS Project was used to schedule the project showing the start and completion dates, locating the critical path(s), showing the sequence and interrelationships between the activities.

- *Step IV: Digitizing AutoCAD drawings to Geographic Information System format*

After creating drawings in AutoCAD, the *.dwg files were transferred to ArcMap a module of ArcGIS. The topologic data structure of the basic design layers in AutoCAD was created as layers in Geographic Information System based on the layers and activities identified in the AutoCAD drawings [73-76].

The Migration of Raw data into GIS format is shown in Figure 12.

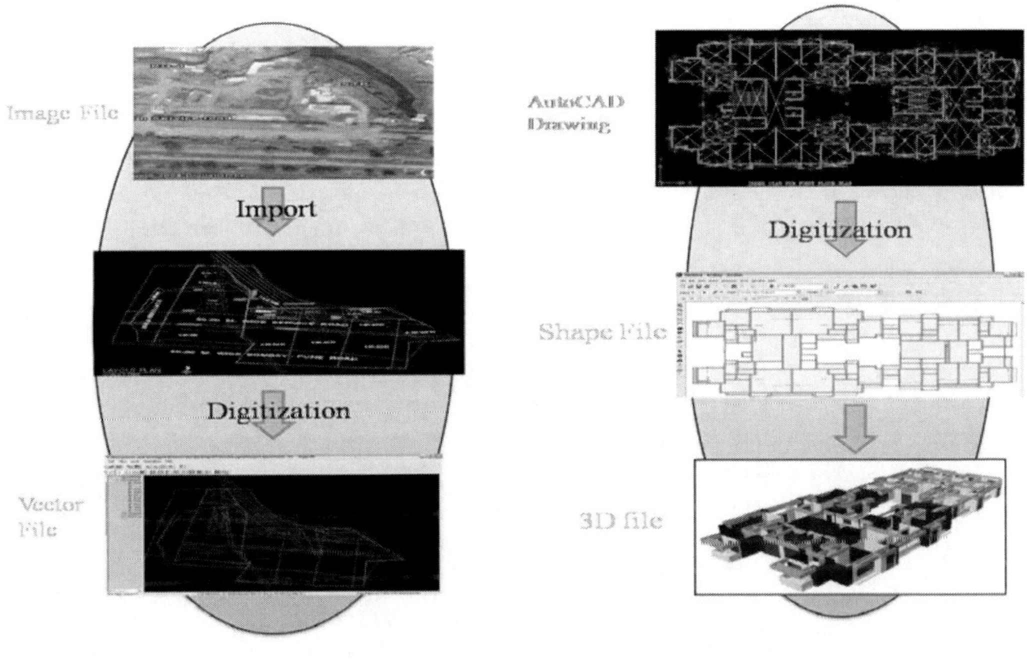

Figure 12. Migration of Raw Data into Geographic Information System Format.

- *Step V: Creating Feature Classes W.R.T. Activities*

The feature classes (Polygons, lines or points) created for activities were merged together into activities defined earlier in MS Project schedule. Thus, the activities, which belong together but are located at different positions, were joined together as one "feature class." For example, all the various components of the Column were merged into one merged feature class called the "BLDGColumn" activity [77-78].

- *Step VI: Creating Database w.r.t. Feature Class, Activities*

The attributes needed for each layer were created in a database.

The database includes information about Z_Factor, TotalVolumeIncum,
Total ReinforcementQtyInkg, StartOfConcreting, PercentageOfConc,
CurrentQtyOfConcreting, StartOfReinforcement, PercentageOfReinforcement,
StartOfShuttering, PercentageOfShuttering, TotalPercentageOfCompletion,

DrawingNo, TotalCostForConcreting, TotalCostForReinforcement, Z_Actual, MoneyUsedForReinforcement, MoneyUsedForConcreting, LastUpdated, Name, ComponentName, Material, BaseHeightInmm, WallPosition for each activity.

- *Step VII: Create updates of MS project schedule*

The user in Step V. enters the percentage of complete information on the activities in the database. This information will be used in Step VII to calculate the percentage of completion for each activity, which is obtained by calculating the average of the percentage of completion of the activities.

- *Step VIII: Run-time application*

The run time application is developed using Visual Studio 5.0 in C# language. With the help of this run-time application a User Interface was developed. Here the user would come to know about the location of the source file. The user interface displays present date and time. It has drop down on the lists for the user to select accordingly.

- *Step IX: Import information about activities and update GIS database*

The percent of complete information is transferred with the help of custom run time application to MS Project at every time when a progress evaluation is made and the application is run. MS Project was run to generate the updated schedule network. The updated schedule shows the progress for all the activities as of the new date of the update (e.g., at the end of every month or daily updates) and the percentage of complete information [79].

Figure 13. 3-D view of the building.

- *Step X: Showing progress of activity in 3-dimensional view*

The project's 3-D view was created in ArcScene a module of ArcGIS. The activity layers created in Step V were converted into 3-D layers in ArcScene. The new layers are shown in the 3-D view created as shown in Figure 13.

- *Step XI: Preparing report as per required format*

The progress of work was shown in graphical format and in different colours. The amount of work done on the various activities could be seen in 3-D view. The project was updated as progress information became available and the corresponding MS project schedule was sent to ArcScene. The updating was done with the help of a custom runtime application.

4.2. Benefits and Limitations of the System

Traditionally, the Construction Project Management schedule does not provide any information pertaining to the spatial aspects or context and complexities of the various components of a construction project. Therefore, to interpret progress information, project members normally look at 2D drawings and conceptually associate components with related activities [80-81].

Different project members may develop inconsistent interpretations of the schedule when reviewing only the Construction Project Management schedule. This causes confusion on many occasions and usually makes effective communication among project participations difficult. This system allows project planners and managers to see in detail the spatial characteristics of the project. All the project members should be able to visually observe the progress, which will help in effective communication of the schedule. The system has to be run periodically over the duration of the project.

This system will benefit project managers, site engineers, and clients in the following manner:

For project manager

- Up-to date information about the progress of work;
- Helps in controlling big project sites;
- Comes to know about the Cost incurred/Spent and the quantity of materials used on site;
- Reduces time for decision making as all information is in one system SITE ENGINEER;
- Controlling the project site by knowing the progress of work;
- Helps in easy decision making for procurement of funds or materials;
- Helps in informing the contractors beforehand about the start of their work;
- Helps in knowing how much more material is required;
- Helps in reducing wastage of materials;
- Helps in ordering the ideal quantity of materials thus by reducing over ordering of materials.

For client

- Helps in knowing the exact status of the project;

- Has a 3-D view of the progress of work thus knowing where large cost has been incurred.

Because of the limitations of this version of Integration of Project Management with Geographic Information System, the interfaces between ArcGIS, AutoCAD, and MS Project, only one database should be available and that is the one which is essential to the operation of the system (i.e., percent completion information), and the system is designed for use in single-user desktop environments. The database of the study area was prepared for columns, walls, beams, slabs, doors, and windows. The database of staircase, site layout, reinforcement detailing, drainage, plumbing, electrical and gas pipelines can also be added for further detailed knowledge of the progress of work. Core software professionals can further improve the scheme of integration of Geographic Information System with project management software. Links of Photographs showing the present status of work can also be added. It should be further explored to integrate Geographic Information System with Enterprise Resource Planning (ERP) analysis, billing software's and System Applications Product (SAP).

4.3. Integration of Remote Sensing Data and System Applications Product in Planning

There are some attempts of use of Remote Sensing and System Applications Product capabilities for planning studies. The core issue in this paper is how to use Remote Sensing and System Applications Product capabilities for the establishment of the data, which will be used for planning studies evolved over the period. This technology briefly introduced and elaborated on the project called "Structural Plan Follow-up Project" in this paper.

A System Applications Product is widely used as tools to digitize Remotely Sensed or cartographic data complemented with various ground-truth data, which are geo-coded using a Global Positioning System (GPS). A System Applications Product can be used to analyse the spatial characteristics of the data over various digital layers. If sequential data are available, quantification of spatial changes becomes possible through overlay analysis. A System Applications Product is an expanding information technology for creating databases with spatial information, which can be applied to both human settlements (e.g., demographic databases) and to the natural environment (e.g., distribution of populations and environmental factors). Most importantly, the combination of both types of database can ensure sustainable management. A System Applications Product will continue to improve as an essential acquisition tool and analysis tool respectively not only in the analytical description of spatial subjects, but also in environmental planning, impact assessment, disaster management and simply monitoring Remote Sensing [82-83].

Global Positioning System, Remote Sensing (including aerial photography), and System Applications Product are technologies that are being increasingly used in the field of data collection, including census mapping operations. Until recently, aerial photography played a major role in geographic data acquisition in urban areas, and the use of satellite imagery, due to its low resolution, was limited to the study of some phenomena like floods and urban pollution. However, recent satellite imagery is extending its use to urban areas with the increase of its resolution to one-meter (i.e. IKONOS satellite), its high positional accuracy, the revisit frequency of only three days, and it's System Applications Product - readiness, despite their relative expensive cost [84]. Coleman [85] proposed a three-dimensional modelling of an image-based System Applications Product to assist in land-use planning and management operations. The model was based on several interrelated databases, including image databases generated from Landsat data, digital line, polygon and point (graphical) files, and a spatial database, which handled the attribute data relating to these spatial coverages. Aerial photography has long been employed in urban analysis [86]. However, this kind of Remote Sensing is still extensively used today while we can benefit from digital image-processing techniques, by digitizing those photographs. The processed satellite image with its ability to provide reliable and accurate data provides excellent possibilities to map monitor and measure the various facets of urban planning and development.

Satellite Remote Sensing, in conjunction with Geographic Information Systems, has been widely applied and been recognized as a powerful and effective tool in analysing land cover/use categories [87-90].

4.3.1. Aims of Studies

Geographic Information Systems and Remote Sensing technologies together nowadays as an integrated technology provides solutions to problems such as urban planning, regional planning, as well as mapping of up-to-date regional and sub-regional road networks, determination of regions as the most suitable for agriculture, industry, residential etc. and regions to be protected due to the fact of environment and so on.

Deputy Ministry for Town Planning (DMTP) of Ministry of Municipalities and Rural Affairs (MOMRA) is very well aware of these transitions and developments mentioned above. Therefore, DMTP is very keen to follow the latest technology and carry out the planning activities accordingly.

As we, all know that planning is a dynamic and continuous process. Due to this particular reason, DMTP decided to start the updating of the digital database mentioned above. The main purpose of that project is to keep the data updated to perform planning studies. The best way to reach that target is to use the integration of Remote Sensing and System Applications Product capabilities for the establishment of data, which will be used for the planning studies in the Ministry.

a. General Objectives of that Project

- Supporting the main planning activities of the cities and towns throughout the Kingdom;
- Integration of planning decisions and information to structural plans;
- Tracking structural plans in short and long terms.

b. Special Objectives of that Project

- Preparing updated structural plans of all KSA Regions;
- If applicable, correcting the errors of the existing structural plans and updating these plans;
- Geographically integrating maps and plans belonging to cities located in KSA regions;
- Preparing technical information and reports related to updating of structural plans consisting of:
  - Population changes;
  - Status of general services;
  - Progress in physical, social, economic and restructuring fields;
- Establishing computerized tracking and updating procedures including storing, retrieval and analysis of data;
- Training the technical staff of the Ministry, DMTP and relevant municipalities and developing in-house capabilities.

4.3.2. Sources and Method

Remote sensing satellite imagery has been used since 1972. The first Remote Sensing satellite was Landsat1 MSS (Multi Spectral Scanner). The sensor technology improved extremely during this period.

Nowadays many information categories for the wide areas can be extracted from these satellite data such as land cover and land use.

Landsat7 ETM has 7 bands varying spatial resolution from 15 meter to 60 meter. One of these bands is 15 meter resolution in panchromatic, one band is 60 meter resolution in thermal channel and others are 30 meter resolution. Detailed information for these bands is given in Table 13. The data-taking period of the satellite is 14 days per location and single frame covers 180km x 180km area on the ground.

Table 13. Landsat7 ETM bands and specifications for each band

Band	**Wavelength *(μm)***	**Resolution *(meter)***	**Spectrum**
Pan	0.45-0.52	15	Visual
1	0.45-0.52	30	Blue-Green
2	0.52-0.60	30	Green
3	0.63-0.69	30	Red
4	0.76-0.90	30	Near IR
5	1.55-1.75	30	Mid IR
6	10.42-12.50	60	Thermal IR
7	2.08-2.35	30	Mid IR

Images, which are taken in spring, are used for the definition of the agricultural and vegetative areas. In some pilot areas, two sets of images are obtained. One set spring images, the other set autumn images. Two sets of imagery give us the chance for more accurate classification and real vegetative areas, while all the images must be taken on cloud free climate.

In order to make geo-registration and rectification of these imaginary, ground control points (GCP's) taken from available geo-registered maps or from Global Positioning System surveys are necessary. Some easy specific points were chosen which could be seen on the image display such as bridge, crossroads, special buildings corners etc. This kind of Ground Control Points locations are entered to the image as a pixel coordinates.

Some sample areas are required to perform the classification which is called training site. These training sites must be taken from different locations and homogenous areas. The classification accuracy directly depends on the number of the training sites. If the number of the training sites increases, classification accuracy will increase also. Optimum training sites determined during the project to increase the accuracy of the study.

The real information for the training sites were taken from field survey and as well as from thematic maps. The information for each training site digitized as polygons on the image.

Two types of classification methods are being used generally. These are supervised and unsupervised classifications. Unsupervised classification is performed for separating the classes according the needs. Unsupervised classification is generally used for determining the main classes and is used as a base for supervised classification. As a summary, unsupervised classification is too much of a generalization and that, the clusters only roughly match with some of the actual classes. In general its value of classification is vital to the spectral content of a scene to aid in making a preliminary interpretation prior for conducting the powerful and good quality supervised classification procedures.

Supervised classification is much more accurate for mapping classes, but depends heavily on the cognition and skills of the image specialist. The strategy is simple: the specialist must recognize conventional classes (real and familiar) or meaningful (but somewhat artificial) classes in a scene from prior knowledge, such as, personal experience with the region, by experience with thematic maps, or by on-site visits. This familiarity allows the specialist to choose and set up discrete classes (thus supervising the selection) and assign them category names [91].

All the Landsat images classified with the supervised classification using the maximum likelihood classification technique to determine land cover, existing land use, and urban areas. Class results performed with maximum likelihood techniques to obtain the homogenous and accurate results. As well as, by using this technique, urban areas can be divided into more subclasses and the vegetation areas in different slope groups can be classified in the same classes.

Unsupervised classification performed for the field study and as well as a guide for supervised classification. Clutters had been separated and represented in different colours with the aid of unsupervised classification. In the field, survey of some special ground characteristic was determined by a hand type spectrometer. The image geometric corrections performed, Image ratio and NDVI (Normalized Difference Vegetation Index) calculated:

NDVI = Near Infrared – Red (Landsat band4 – band3)) Near Infrared + Red (Landsat band4 – band3)).

All the classification results controlled with a second field study to obtain the accuracy of the final data. If the results do not meet the needs after the second field study, the classification edited and reclassification applied. Result output raster images converted into ESRI Grid files to prepare the thematic maps in Arc View and/or ArcInfo platforms. Final data converted into geodatabase, Arc View shape files and/or ArcInfo coverage.

After converting the data obtained from Remote Sensing method to our geodatabase as a layer mentioned above, we have that opportunity to make analysis, decisions, and follow-up in our planning studies with the data stored in our geodatabase like maps, graphic and non-graphic data, and tables. The classification of the satellite images, extracted the required information for land use and land cover with the geo-referenced graphic data is a part of data processing. At the same like existing / proposed road networks, subdivision plans, existing / proposed land use areas from the structural plans etc. are the features of the investigated area. In this case, collected data becomes much more intelligent as a planner in DMTP that can now make better decisions on the structural plans by superimposing both information.

4.3.3. Outcomes

Having enough quantitative and qualitative information concerning the natural resources of KSA, we can make the most convenient planning. In order to reach that aim we must integrate the Remote Sensing technologies with Geographic Information System. By taking satellite images from different dates and times, we can discover the land use situation yearly from the current situation to 10-20 years back and this will help us to follow-up the change in the land use of the related regions.

With the help of that database we can produce new land use structural plans like determining the new airport locations, recreational areas, solid waste areas etc. Besides, dealing with the agricultural lands we can determine the agricultural policy of the country by taking into consideration the under-ground water storage areas plus by determining the menacing urban or industrial development through agricultural lands, governors can be informed and the development can be directed from productive plains to other possible locations.

There is a lack of geographic data in Middle East countries. For that reason and with the consideration of time and cost of any planning project, it is wise to utilize a Geographic Information System technology integrated with the Remote Sensing to establish geo-referenced digital data. Spatially integrated raster data allows it to be viewed, queried, and analysed with other geographic data. There is clearly a good scope for urban Geographic Information System development in Saudi Arabia and a lot will depend on how Government of Saudi Arabia and funding agencies are able to come up in near future. This project, presented in that article, is a successful example and shows a simple but powerful solution to developing and even developed countries lacking up-to-date geographic data.

4.4. ENGINEERING AND ENGINEERING MANAGEMENT SPACE TECHNOLOGY APPLICATION

Up to date circumstances of any engineering activities embrace a wide area of activities and involvement of a large engineering features' solutions. The use of different instruments and tools of engineering is demanded for meeting requirements of the standards and regulations required for a project execution. It is being used in engineering from the stage of tender package developments up to the project execution and as built document archiving. The content of such developments can be included the spatial (layouts, drawings) and non-spatial (specifications, cost estimates etc.) aspects of information in a construction project.

With regard to this approach, it is necessary to collect required information in the early stage of engineering activities. It is obvious that an accuracy and detailed data will create a good environment for the success of the engineering works in any stage of implementations.

It became easy and useful to use the advance of computer technology to integrate the recent advances of the high technology in wide areas of engineering applications. In this integration, the drawings can be drafted using a computer – aided drafting (CAD) program (AutoCAD) and construction schedule would be prepared using Project Management Software.

In the meantime, the use of the space technology advances opens a large success in engineering developments, which is a key instrument in project implementation and management processes.

Engineering in Industry has a huge number of implementations beginning from different types of engineering disciplines involvement up to all required documentations like drawings as built layouts and financial issues and legislations. At the same time, it is necessary to monitor progress of the project executions permanently and add all available data into the appropriate achieved folder for the future.

There is no doubt that the collection of required information is highly important before any engineering activities intended to construct any engineering facilities. It has to be considered all engineering features on site, which makes available successfully integrate existing and forthcoming communication and other necessary engineering structures into a unit system. It is especially valuable action when the volume and scale of project is big enough.

At the same time, any engineering activities on site demand execution of control of works. There is pressure on the project managers to shorten the delivery times and thus the current scheduling and progress-reporting practices are in need of substantial improvements in quality and efficiency. Application of methods Remote Sensing with further development of Geographic Information System can be successfully applied for elimination of such disadvantages. A Geographic Information System and project management software with visualization were recognized as one of the most important tools for achieving this goal.

A Geographic Information System is not a tool for presentation of data. A Geographic Information System does in fact, create high quality maps that communicate considerable amounts of information in an efficient and attention-getting manner. A Geographic Information System is both a database system with specific capabilities for spatially referenced data, as well as a set of operations for working with the data.

4.4.1. Remote Sensing and Geographic Information System Applications

Geographical information System is an instrument for integration of appropriate data into the geographical system. The data can be provided by using Remote Sensing methods throughout satellite images. This instrument allows gathering, storing, analysing, editing, sharing information and successfully using in wide required areas. In the meantime, Geographic Information System is a tool creating an environment for interactive quires, analyse the spatial information, edit data, developed maps, and present outcomes of all these operations.

A Geographic Information System technology is being used in diverse application areas related civil engineering, such as:

- engineering management;
- land survey management systems;
- environmental problems;
- hydrologic systems development;
- infrastructure management systems;
- irrigation and drainage system design and construction;
- transportation; and
- Other areas of engineering management.

Some applications of Geographic Information System related construction can be found out in the literature [92-93] where has used a GIS technology to construction contractor prequalification. In the [94] where demonstrated development and integration of Geographic Information System aids managers during their activities related to onsite soil condition. In the work [95] presented the system, Material Plan, which integrated Geographic Information System - based on cost estimates with construction scheduling to automate planning tasks required for materials layout. The studies of [96] investigated the roles of internet - based Geographic Information System in e - commerce systems and used of Geographic Information System capability to integrate, to manage, and to analyse spatial information. There is integrated Geographic Information System with visualization techniques for planning and visualizing of dam construction processes in the [97].

The main goals of presented paper are following:

i. Remote sensing and Geographic Information System as a tool data collection in project design;
ii. Remote sensing and Geographic Information System in project management processes; and
iii. Remote sensing and Geographic Information System application in engineering facility construction.

4.4.2. Advance of Technologies Application

Today the use of Remote Sensing methods and Geographic Information System technology is one of the interesting computer - based technologies and it is the initial stage of application of their potential and capability in engineering. This technology supports the interaction of multiple participants such that they can approach problems in a more comprehensive and systematic way. One of the main benefits expected from the application of Geographic Information System for contractor prequalification is the creation of a comprehensive database. This database can provide a wide range of construction users with a mechanism for rapid retrieval and manipulation capability to satisfy their need of spatial and descriptive information required in the process [98].

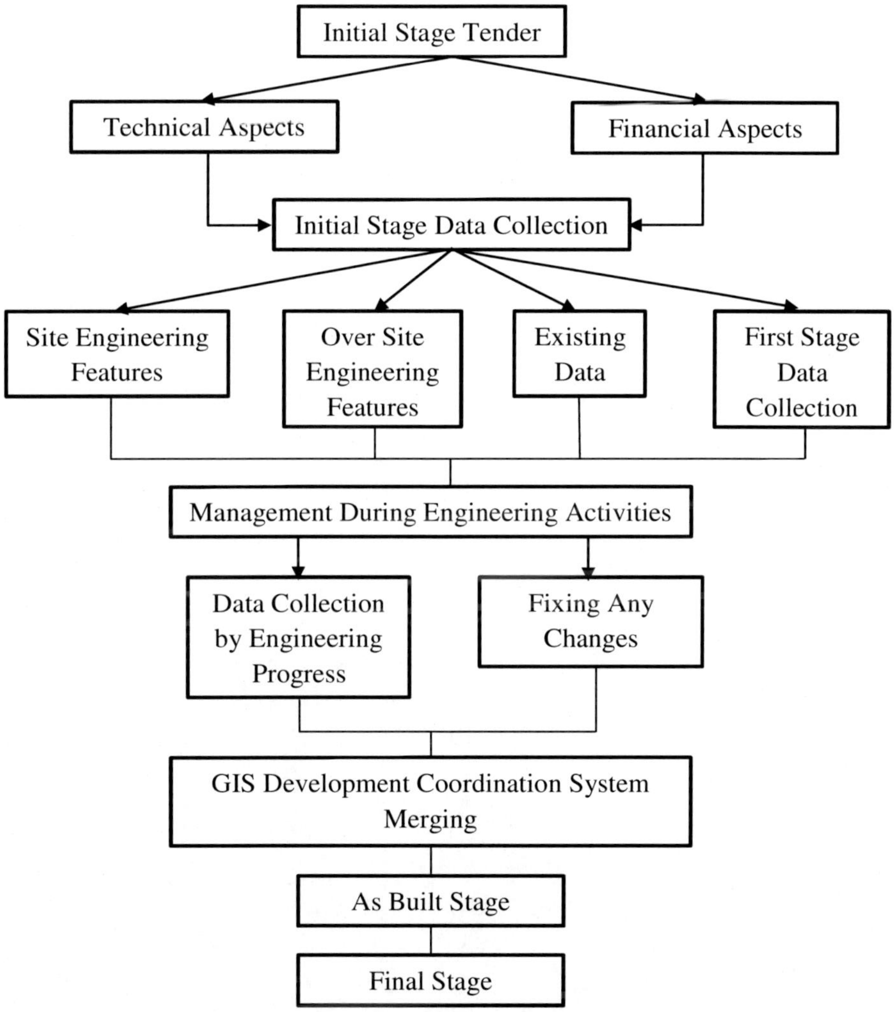

Figure 14. Description of the process of development of project stages by means of Remote Sensing and Geographic Information System technologies.

Specialists in the engineering always look for a tool for collection of appropriate, right data and efficient storage for easy access in needed case, analysis, and presentation of geographic data. These efforts have apparently been the result of increasing demands by users for the data and information of a spatial nature. For solution indicated problems, the use of Remote Sensing and Geographic Information System developments is an excellent instrument. This approach of using such a system goes beyond the description where analysis, modelling, and prediction are included. Remote Sensing and Geographic Information System are the means contenting organized collection of computer hardware, application software, geographic data, and personnel designed to efficiently capture, store, update, manipulate, analyse, and display all forms of geographic referenced information.

The structure of the Remote Sensing and Geographic Information System can be described as the chart reflected below in Figure 14.

It is describing the stage of processes from the initial point of tender package developments and project implementation itself. There are two main levels of data collection and processing. The first level is related to the tender phase and the second is the before starting of the engineering activities data collection, processing and analysing.

The next level is the most complicated part of engineering developments. It is necessary to monitor all the process and progress of engineering activities during project implementation phase. A suitability and efficiency of the quality, human resources, timing and financial management issues will depend on how the monitoring of engineering works will develop.

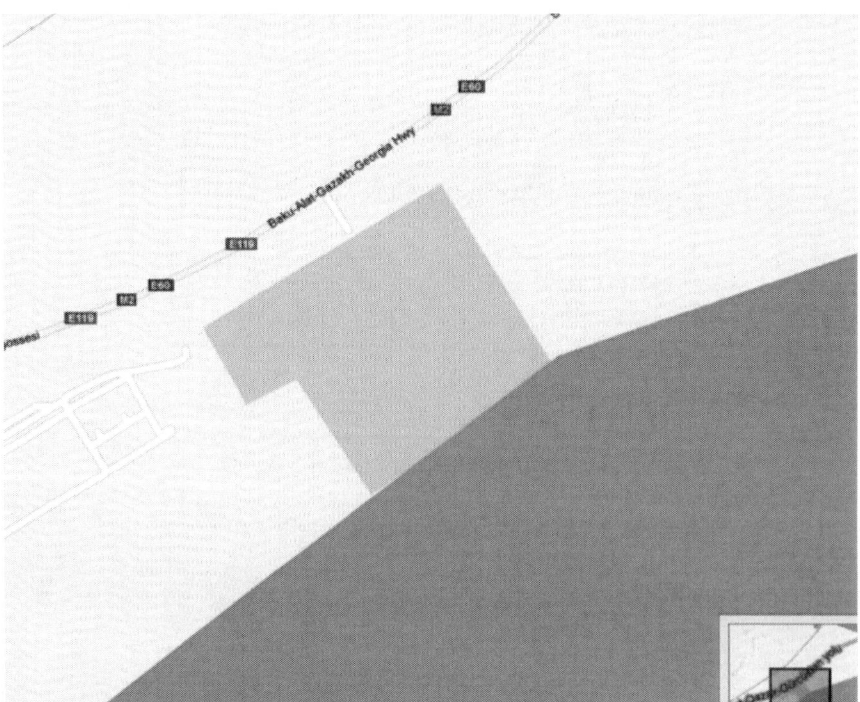

Figure 15. Layout of the selected industrial site.

4.4.3. Remote Sensing and Geographic Information System in Engineering Facilities Design and Construction Processes

It is required to evaluate existing condition of the site before any engineering works where the execution of industrial facilities is intended. The sketch in Figure 15 demonstrates empty site with a lack of engineering features. It is the site where the construction one of the important plant of Azerbaijan not far from the Baku city was planned.

This sketch was integrated into the space image with a high spatial resolution 1m. It is desirable to summarize as more as possible data into this system for further processing. As it shows in Figure 16, the site contains almost a few engineering elements.

Figure 16. High spatial resolution space image for 2009.

An investigation based on space data by means of Remote Sensing methods and developed Geographic Information System technology opens an opportunity comparatively easy to integrate existing engineering facilities into the system-constructed structure.

Figure 17 shows the actual conduction of the site after almost completed plant with the main engineering facilities.

Figure 17a. IKONOS space image processed with engineering features of industrial site.

All involved engineering elements to this data are reflected in Figure 17b as legend.

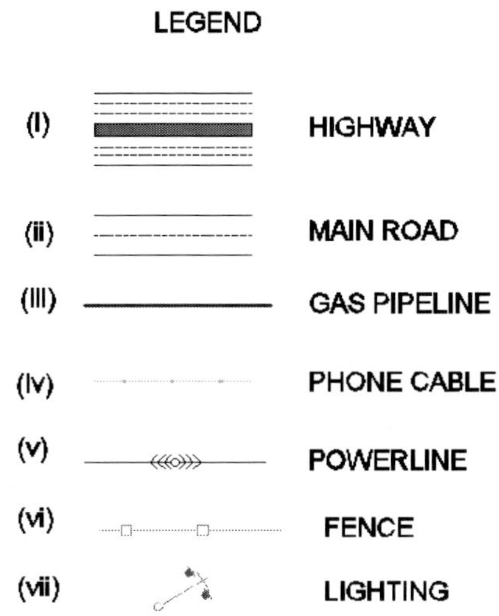

Figure 17b. Legend of processed space data.

From this view, the advantages and comforts of merging data with the use of Remote Sensing methods and Geographic Information System technology applications can be easily discovered.

An analysing the above-mentioned aspects, it can be concluded with the following results and outcomes:

- The increase in classified layers of the developed Geographic Information System based on Remote Sensing methods will create an environment of comprehensive application of data within this system.
- Remote Sensing and Geographic Information System technology has an advantage for engineering management in engineering planning, time, human and financial management issues. A Geographic Information System can provide powerful information – not just how things are, but how they will be in the future basing on changes.
- Remote Sensing and Geographic Information System usage for engineering management system given as an example demonstrates the flexibility of the use of Geographic Information System technology in the wide area. Integration of collected information with needed other investigated or monitored fields can be successfully adopted into common targets for further developments.

4.5. Space Technology Applications – Cost Effectiveness

Space science and technology application has a significant place in our life. For this purpose many countries spend big financial resources for research and exploration aims for sustainable economic and political developments. It gives attention to the areas as how to best organize research activities, how to best use, estimate its utility and profit (real and potential), how to best increase and manage efficiency, how to best identify new ideas and innovations, how to properly fund research works and new ideas and innovations, and how to correctly evaluate results and outcomes. There is no doubt that all these mentioned issues are important for successful planning and management of the space science and technology researches and explorations, starting new launch systems, data collection and data processing processes.

4.5.1. General Approach

It is necessary to mention that there is major aspects and problems on offers of innovations in management, organization, process evaluation and estimation which suggest a new research efficiency criteria, new approach of development, new methods for assessments of new ideas, innovations in space science and technology as well as new methods in patenting technology. At that time, the assessment and evaluation of research outcomes and results, research efficiency criteria, innovation in organizing of scientific research and explorations become necessary [99].

Many powerful countries like the United States of America and Eastern European countries spend a big amount of fund on research works for development and improvement existing capacity and achievements of a suitable use of the space science and technology outcomes in our daily needs. We are in a stage of thinking about a cost efficiency of the space technology application as far as the cost in enough high of the area.

4.5.2. What Is Required for the Success?

It is obvious that any country involved for use of advances of space science and technology is necessary to have an endeavor for the area. It should be a first step for any undertaking. From this step is starting the use of offers to create a positive environment for research and exploration how to increase an efficiency in the space science and technology.

The next very important side of high progress in efficiency of scientific research and developments in the space science and technology is the efficient use and management of available funding resources. It is obvious that financing of any project is always limited everywhere. Unlimited funding is inconceivable. For the progress, the development and creating of the right organization of scientific funding and research areas is necessary and important. It is required to consider the best organization and identification ideas and innovations for research works. The selection of the right people who will be more effectively implement a project with the expected outcomes is also priority of the issue.

It is not a simple task for investigations of all indicated above matters. However there is a way of the problem solving based on the use of the method of the counter management with target and aim definition. This method gives opportunity to use the feedbacks of all the processes for success and to develop common criteria that help to solve the problems of selection, management and organization in order to save money and achieve practical success.

4.5.3. What We Have to Do?

The support of new innovation based on the concept and an idea in the space science and technology advances. It is very important to clarify how the innovation with reflection of concept and idea in appropriate area will be selected.

The second issue is the nomination of the right researches for the projects that create environment for a high efficiency access within the framework of project implementation.

The first approach of the progress in innovation can be undertaken with the following description. It can be intended to be considered the five stages of innovation developments:

The first stage of announcement of grants, for instance, scientific grants for opportunity on inquiring of innovations based on proposed new concepts and ideas. In this stage it is required to touch the sensitive areas of the space science and technology and applications which need more attention for development. Collection and gathering of purposeful data in the selected area of research and development are the main reasons of future success in the expected explorations.

The second stage of innovation progress is the embracing of a new approach of concepts or idea sharing. In this stage to find out a new concept and idea based on actual research outcomes initially done and practically used for some purposes is important. The criteria should be developed for the stage with undertaking indicated main aspects of innovation for its further progress.

The third stage reflects inquires related to experience and knowledge in the space science and technology developments. It can be evidence as published scientific papers or other publication sources. Conference proceedings and materials are also good promoting sources for evaluation and decision making. This stage is mainly directed for validation of theoretical capacity and achievements of potentially granted researches and specialists.

Within the scale of the fourth stage of evaluation in any innovation processes the consideration of available applications of outcomes in the space science technology research and development as an initial stage of status identification is highly desirable. These created criteria can provide a clarification of the challenge in the area of making decision of selection for forthcoming project implementation under the umbrella of grant offers.

Looking for the future. It is the last stage of innovation approach. It has to be taken in the mind for the perspective implementations. Sometimes some innovation is not ready for realization for the time being. It needs some technological or technical developments for carrying out concepts and ideas of suggested innovations. Of course, it is very difficult to evaluate such suggestions where impact of consequences is huge. Undoubtedly, it makes difficult to consider those circumstances during evaluation process. It is obvious that in an initial stage such concepts will have a less efficiency than existing innovations. It should be understand that in the future innovation efficiency will become significantly more than existing concepts and ideas. It is vision for the future which also demanding of funding in this stage.

4.5.4. Pattern of Application

In conformity with general approaches consideration of a new innovation efficiency E can be described as expected possible profit, P, divided by cost, C:

$$E = P/C \quad (1)$$

This approach needed to be taken into mind in any issues of the space science and technology research and developments innovations.

There is a project implentation which has been granted by the University of Wisconsin–Madison Disater Management Centre, USA for the sphere of "Application of Remote Sensing and GIS technology to reduce flood risk" executed under ProVention Research & Action Grants.

The main aims of the grant were following actions and stage:

(i) investigation of the main stakeholders and establishment of the contact with appropriate authorities in the local scale as well as head office that would be interested and provide a contribution in the project progress;
(ii) area investigation for further space science and technology applications;
(iii) an access to the needed space data.

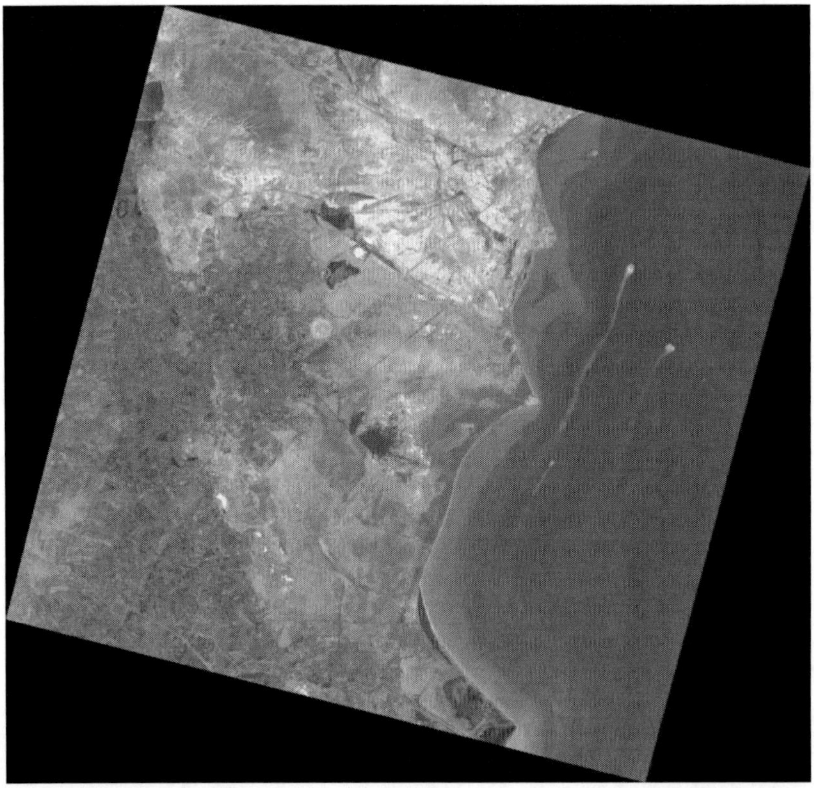

Figure 18. ALOS satellite image.

It is necessary to emphasize that one more way of the high efficiency of innovations is to use the outcomes of the project sources for a multiply purposes. For instance, selection of the space image has a big importance in this development. It has been selected onshore and offshore areas for space image selection within the frame of the above mentioned project (Figure 18). It makes available to enhance an advantage of the data sources for the reason not only for flood investigations but also for environmental, ecological problems of the Caspian Sea, for instance, costal line, oil pollution investigations and so on [100-101].

The outcomes of the project have reflected indirectly climate change problems linked with the impact of river flood during the study of the reason of the natural disaster. This information was very important for authorities for decision making [102].

An efficiency of the used innovation during project implementation was very high which was containing the reduce of river flood impact providing protection works for minimization carried out by local authorities in the place.

This paper describes the general approaches of innovations based on new ideas and concepts for reaching of high efficiency. An application of advances of the space science and technology in research and developments was used for this purpose as an example.

The aspects such as how to develop system approach and criteria to achieve a high efficiency in innovations intended to be used in the future applications and implementations are shared in this paper.

CONCLUSION

The use of the space science and technology advances is very useful tool in the engineering management process. It consists of the integration of a large number of data with the system as a land management system containing wide engineering features with elements of environmental issues. It is obvious that demonstrated in the Chapter topics express importance and opportunities of application of Remote Sensing and Geographic Information System in an engineering management process as long as information in any stage of engineering activities needs the successful management approach.

Chapter 5

SPACE TECHNOLOGY: CIVIL PLANNING

ABSTRACT

This Chapter captures Remote Sensing and Geographic Information System application in civil planning. It contains the areas of design and construction processes of civil planning based on Remote Sensing and Geographic Information System applications.

It is a tool for achieving the targets in civil planning as a key opportunity of collecting necessary information successfully and of integration into the space data received by Remote Sensing method. An advantage of the use of such a tool makes available to manage all sources (human, technical, technological, financial, and time) in any stage of project execution.

Keywords: Remote Sensing, space image, data processing, civil planning, data collection and data processing, field measurements

INTRODUCTION

Space technology takes a valuable position in wide areas of engineering. Up-to-date space technology advances find a good place in civil planning. Huge articles and scientific papers have been dedicated to this area. It reflects the importance and necessity of the space technology application in indicated engineering discipline.

It is not far the day, which space satellite constellation would be directly implemented for engineering needs launching those with such a purpose. The use of the space technology cannot be limited for some scale. It can be used in social, public purposes in particular industrial spheres.

Remote Sensing and Geographic Information System create a convenient environment for evaluation process of the engineering processes, which are very important for challenging of carried out works. Remote Sensing and Geographic Information System as tools make access of data to link engineering and non-engineering activities as a whole, especially in civil planning which contains heavy executions.

5.1. Remote Sensing and Geographic Information System Overview in Different Spheres

Remote Sensing observations provide data on the Earth's natural resources in a spatial format. The Remote Sensing data has the advantage of synoptic view and large area coverage. There is one more advantage is an ability to collect a data with a high accuracy with simplify access to the needed information source. The information required in the field of civil engineering is derived mainly from analysis of image patterns presented in the data. These patterns reflect the influence of the type of parent material, geological processes undergone, the climatic, biotic, and physiographic environment and man's activity. Thus, applications of Remote Sensing in engineering involve the recognition of basic landforms as indicated by the pattern elements on the image. It is necessary to add that space image patters can be processed depending on the nature of the task required for engineering executions.

Spatial data Collected by method of Remote Sensing can be efficiently handled using Geographic Information System - a tool that allows synergism of map data and tabular data. A Geographic Information Systems also allows the integration of these data sets for deriving meaningful information and outputting the information derivatives in a map format or tabular format [103].

In a variety of areas of engineering projects, Remote Sensing and Geographic Information Systems techniques can become potential and indispensable tools. A large number of engineering application spheres include regional planning and site investigation, terrain mapping and analysis, water resources engineering, town planning and urban infrastructure development, transportation network analysis, landslide analysis, etc.

5.1.1. Collection of Planning and Site Data in Space Data Processing

There are executions for engineering activities required to be done before, during and after project implementations. Site investigations, in general, require topographic and geologic considerations as the first stage of engineering developments. Remote Sensing data permits such assessment. In case of dam site investigation, information on topography is essential. Geological consideration involves the different soil and rock types and physical properties.

In selecting river-crossing sites for bridges and pipelines, an important consideration is the stability of slopes leading down to and up from the water crossing. Such slopes include riverbanks, terrace faces, and valley wall. History of river erosion and sedimentation would give clues needed for locating the sites where scour is likely to occur. High spatial resolution satellite data with stereo vision capability can facilitate depth perception in the above said investigations and for regional planning of large commercial airports, harbours, industrial towns, and recreational sites. The hydro geological and geomorphologic information along with geological structures derived from satellite data are very useful in sitting the ground – water bore holes.

5.1.2. Terrain Mapping and Analysis

Assessment of the performance of the terrain for specific developmental activities can be made through terrain evaluation. For this, terrain information can be acquired from RS data and by generating the Digital Terrain Model (DTM). A DTM is an ordered array of numbers representing the spatial distribution of terrain characteristics stored in a computer to enable the determination of any quantitative data pertaining to terrain. DTMS is useful in investigation of a number of alternative horizontal and vertical alignments of canals, roads, pipelines, or corridors for any such applications. In engineering construction like dam, the knowledge of material, comprising the terrain is essential for proper planning, location, construction, and maintenance of engineering facilities. For computation of hydrograph parameters like peak runoff rate, time of concentration and time to peak, the height and slope information derived from Digital Elevation Model (DEM) are useful. The information on regional engineering soils is essential for general planning and site evaluation purposes [104].

High spatial resolution satellite data can be analysed for delineating various landforms, mapping of soil classes contributed to engineering construction, delineation of landforms – engineering, soil relationships and grouping of landforms with various physiographic setting or terrain associations. Remote Sensing based inventory of construction material such as boulders, quarry rock, sand – clay mixtures etc., would help to locate suitable sites of construction materials for building up the water resources engineering projects like dams, bridges, etc., across the rivers.

5.2. Water Resources Engineering

The water storage built in through reservoirs, tanks, etc., is often reduced due to sedimentation. Remotely sensed data can be used to monitor the water bodies over time and assess the silting condition. In case of gauged reservoirs of medium to large sizes Remote Sensing data can provide an assessment of sediment volume and reduction in the capacity of the reservoir. In case of small water bodies such as tanks, it is possible to come out with a list of problematic tanks with symptoms of heavy siltation and loss of water holding capacity. The condition of tank bunds, fore shore encroachment, etc., also can be analysed with the help of high spatial resolution Remote Sensing data.

Characterization of water bodies in terms of geological, geomorphological, hydrogeological, soil and land use / land cover parameters carried out using Remote Sensing data enables conservation of land and water resources. The Remote Sensing based on input integrated with ground based information through Geographic Information Systems is useful for broad reconnaissance level interpretation of land capability, irrigation suitability, potential land use, water harvesting areas, monitoring the effects of soil and water conservation measures, estimation of run off and sediment yields and monitoring land use change including land degradation. The commercial areas of irrigation projects are fed by different sources of irrigation like reservoir, tanks, ponds, and wells. Assessment of command areas and crops would be highly useful in water release policy or conjunctive use of water in the overall project command area. Satellite data has been advantageously used to obtain such information on surface irrigation projects. In case of floods, appropriate flood management work has to be executed to reduce the damages and utilize the floodwaters. Using multi-date pre-flood and post flood images damage due to flood can be successfully assessed. This will help in mitigation and relief distribution. Satellite derived snow cover assessment is being extensively used as an important in putting snow melt runoff prediction models for assisting the multi-purpose reservoir operations.

By analysing multi date Remote Sensing data, it would be possible to monitor the effects of dam construction. Remotely sensed data of pre and post dam construction can reveal the forest and other land at different water levels. This would also help in preliminary investigation of finding suitable areas for human resettlement.

A study of the feasibility of inter basin transfer of surplus flood flows, Remote Sensing data can be cost effective. In large area reconnaissance studies various technically feasible and economically viable alternatives can be arrived in locating surplus flow diversion routes to water deficient basins. In addition, reservoir sites to store the surplus flows in these basins could be identified. Such projects of large dimensions require considerations of land use / land cover, soil and geological mapping, terrain evaluation, construction material inventory etc. the latter are derived from satellite Remote Sensing data of particular resolution depending upon the scale on which such information is required.

The water storage built in through reservoirs, tanks, etc., are often reduced due to sedimentation. Remotely Sensed data can be used to monitor the water bodies over time and assess the silting condition. The condition of tank bunds, fore shore encroachment, etc., also can be analysed with the help of high spatial resolution RS data.

5.3. Town Planning and Urban Development

The unprecedented growth of urbanization is one of the main reasons of the rising of problems of housing, sanitation, power, water supply, disposal of effluents and environmental pollution. Systematic mapping and periodic monitoring of urban land use is therefore necessary for proper planning, management and policy making (with the help of Remote Sensing and Geographic Information Systems optimal master plan for development and management of urban settlements can be prepared). There is no doubt that information accepted by the use of space technology advances is one of the best sources for decision-makers as the data with high accuracy and sufficient covering required information for right conclusion of existing weaknesses and problems [105-106].

For sustainable development of urban agglomeration, optimal urban land use plans and resources development models need to be generated by integrating the information on natural resources, demographic and socio – economic data in a Geographic Information Systems domain with the integration currently available satellite data [107].

5.4. Transportation Developments

With the help of high spatial resolution data, mapping of road and rail network can be accomplished. This facilitates in deciding optimal routing for transport of construction materials. Even, the identification of village roads is possible in certain cases.

Highways are parts of the infrastructure that makes up the spinal cord of modern society. A Geographic Information Systems can be successfully used as a valuable tool in the process of planning and design of highways process to obtain an optimum highway route alignment, which is economical, suitable and compatible with the environment. The various types of data have to be considered simultaneously.

Handling and managing this large amount of data manually, is not easy. For these purposes Geographic Information Systems comes to help, because of its inherent property of handling large bulk of spatial data, non-spatial data, and its analysis. Remote Sensing images of the study area are used as the source (spatial data). Various collateral data from various offices are collected in order to be used as non-spatial data. These images are used to prepare the digitized formats required for the Geographic Information Systems techniques [108]. Using the Resistance concept (such as areas suitable for the new alignment were assigned a low resistance value, whereas the areas not suitable for the new alignment assigns a high resistance value) the data is prepared for analysis. In general, Spatial Analyst tool of ArcGIS version 8.1 uses for performing the analysis.

There are ample evidences of applying the recent advances in satellite based Remote Sensing and Geographic Information System technology in various fields of civil engineering. India's space program ensuring continuous availability of Remote Sensing data and launching the future satellites carrying high spatial and spectral resolution sensors can go a long way in providing useful information required for civil engineering applications.

5.5. Remote Sensing Methods and Geographic Information Systems Technology Application in 3-D Performing

It is necessary to indicate that geographic information systems have existed for a long time, but they were not called geographic information systems then. They were called geographic databases or something like that. Some of the earliest works in geographic database development have been started to be named as Geographic Information Systems in the early 1970s.

Actually, when LANDSAT satellite went up in the early '70s, it became a principal source of geographic information. Up to that time, maps were digitized. The U.S. Geological Survey had 7 1/2 minute quadrangles (1:24,000 scale) and 1-degree quadrangles at a scale of 1:250,000. The source of the data for those maps--elevation and features (roads, streams, etc.) came from airborne flights. When the Earth resources satellites went up, they became a major source of information that not only showed where buildings, office parks, and forests were, but also helped to get a feel for exactly what was on the land, such as croplands, grass for pasture, and even individual buildings. It became a way to gather what is called land cover information and land use. Therefore, in the middle to late '70s, that became a principal source of data that went into Geographic Information Systems [109].

5.5.1. Satellite Data and Processing Technology

Based on the enhancement of technology achievements it was the focus of a company named ERDAS (Earth Resources Data Analysis Systems), which was a spinoff of the group, at Georgia Tech, USA. It used Geographic Information Systems and some of the same techniques that Harvard had developed but it integrated that type of geographic data with imagery. Therefore, ERDAS was one of the first companies integrating the use of imagery and geographic information systems data where data was integrated into coordinate system.

Parallel with the development of IMGRID, another branch of Geographic Information Systems was starting to be developed that really did not have anything to do with imagery. It was taking maps, digitizing roads, and doing everything in polygons rather than in grids. There were a whole other set of software and techniques developed for polygon overlay in this polygon Geographic Information Systems.

It would have any kind of arbitrary polygonal boundary and inside that boundary can exist different attributes. It contains many acres. It is a field and cotton is growing on it.' That polygon is a set of vectors, nodes, and links between the nodes. It is an enclosed area with attributes attached in it. On a grid, on the other hand, the same data can be represented, but every point has an attribute that says what it is and each point has exactly the same area.

It can be explained in detail that in an image there is a pixel size. In a satellite image, the pixel size from LANDSAT is currently 30 m. Therefore, you have an image, and each point on the image is 30 x 30 m. A lake, in an image, may be a set of a hundred pixels that are all considered lake. A polygon, on the other hand, would be just the outside boundary of that lake with an attribute that would say it is a lake in between. It is a different representation of data. It turns out that there is a lot of detail in imagery that would take too long to put into polygons. A verification of imagery information in Geographic Information Systems development in some cases is necessary with the integration of polygon data during space data processing stage. Honestly, it is important to generalize the information in the imagery to put it into a polygonal Geographic Information Systems.

Obviously, there is always a lot of work going on in the military in terms of terrain data. In the mid-'70s, actually it has been done many works with the Army Engineers Topographic Laboratory. Their job was to create terrain data and maps for the military, mainly the army. The Defence Mapping Agency was also building these maps. In the '70s and '80s, it has been decided to switch from paper maps and go into digital maps. In case of existence of digital maps, it can be wind up creating geographic information systems data. The geographic information system is essentially a way to present that map data.

The military started using imagery a long time ago. They were instrumental in making sure the imagery came in geometrically corrected to the map of the same area so that it could look at how things changed over time.

The early LANDSAT, launched in '72, had a resolution of 79 x 57 m. So, it was no square pixel. The LANDSAT in the early '80s had a new sensor called the Thematic Mapper. It had a pixel size of 30 x 30 m. SPOT is a French satellite and it has two different types of systems. One is a colour infrared digital camera that has three bands: a green, red, and near-infrared band. Its resolution is 20 x 20 m.

On the same satellite, there is another sensor, which creates panchromatic information like a black-and-white photograph. Its resolution is 10 0 x 10 m. Each sensor has different pixel size, but it also has different colour information associated with it.

There are about four companies in the U.S. that have permission to launch satellites with as well as 1-m resolution. The data from IKONOS is one of the common used source for image satisfaction where high-resolution images required. In addition to that, it is necessary to differentiate those private companies such as Lockheed and Space Imaging launching their own satellites with resolution. Some of them are multispectral and some are panchromatic. Panchromatic is black and white images. Multispectral is colour images that makes possible to get more colour information. The black and white image is fine for showing position of things but it is hard to interpret the vegetation health and things like that for remote-sensing data. It creates obstacles in the space image processing for data classification.

General approach of use 1m resolution data is going to be mapping for counties or even individuals on a land-parcel basis. It is to hit a market that is dominated by these mom-and-pop air photo firms that go out and take aerial images to create detailed maps of a subdivision or other small areas. It is not really the traditional Geographic Information Systems market, which is more natural resources oriented, where it can be used vegetation analysis and classification or crop estimation. That is too small market. For instance, the military has its own set of satellites, but many times, it is easier to get commercial data and actually use it. In addition, for countries that do not have their own surveillance satellites, this would be an obvious way to do surveillance for space image data access satisfaction.

One of the innovations that have come out comparatively recently is the merger of three technologies. Those technologies are remote sensing, Geographic Information Systems, and computer graphic visualization. The remote sensing and Geographic Information Systems merged in the last decade with ERDAS and INTEGRAPH and a few other companies that used imagery behind their vector and polygon data. Today it is very difficult to separate integration of remote sensing and Geographic Information Systems technology in wide areas of human life.

5.5.2. 3-D Image Performing

The ability to create a 3-D perspective scene has always been attractive because of our life in a 3-D world reflection. If we want to understand geographic data, it makes sense to visualize it in 3-D performance creating an excellent environment for clear understanding and processing of features. The problem before was the computer systems of low capacity limiting the use and application of properly expectation in processing actions. They were too slow. It used to take hours to generate one perspective scene. Now with all the computing advances, the Pentium chip and graphic computers such as those from Silicon Graphics and Evans and Sutherland, there is a level of computing power available on a workstation that allows creating 3-D scenes in real time. It is essentially a work station-based flight simulator. By generating those scenes in real time, it can query and analyse the Geographic Information Systems. View sheds (an analogy to watersheds) have always been an important part of what the military is doing in terms of terrain analysis. However, Geographic Information Systems people never really considered view sheds as part of their analysis capability.

The military always had two types of use for geographic information systems. One was called terrain analysis, which was line-of- sight view from the top of things. It got into fields of fire and things like that.

The other use for Geographic Information Systems had to do with mobility. It analysed the terrain. What is it going to take to move a tank from here to there, based on the soils, vegetation, roads, and stuff like that?

Those were always two separate types of analyses. The civilian Geographic Information Systems never really did terrain analysis. However, when it needed to be started to include the visualization in the Geographic Information Systems, the terrain analysis rather came along free.

Actually, Remote Sensing and Geographic Information Systems are separate actions in data collection and processing. It is important to mention that the integration of Remote Sensing data into Geographic Information Systems technology gives an advantage of comprehensively problem solving. Remote Sensing data is involved in Geographic Information Systems as a part of the whole process of data processing. Remote Sensing provides land cover types of data for a Geographic Information Systems. A Geographic Information Systems is a set of layers of information. It could be elevation data, soils data, population data, geology data, hydrology data, or vegetation data. A Geographic Information Systems have many more different layers of information for one geographic area than remote sensing. Remote Sensing gives information used to create one layer in a Geographic Information Systems, which is land cover. An exception point of remote sensing integrated into Geographic Information Systems is the geographically location of all data into coordinate system.

Remote sensing provides images in multiple spectral bands. Based on image processing as pattern recognition it can look at those images and determine of features on the Earth.

5.5.3. Data Processing Classification

Space image consists different spectral characteristics that are some of the remote sensing bands allow differentiation between different rock types. There is different spectral response between, say, acidic (e.g., granite, gneiss) and basaltic rocks (e.g., olivine). Some of the spectral bands, for example, in LANDSAT can pick that out. In any case, each of those spectral bands is an image. A person can look at it and interpret it. That image is not information until that person interprets it.

There are some other opportunities in sensor developments installing in the satellite, which is a hyperspectral sensor. Instead of one to three spectral bands, it is going to have hundreds of spectral bands. Each bandwidth is 10 nm, real small slices. If we look at a spectral curve, it is essentially almost get a continuous curve.

The problem with the LANDSAT bands and the SPOT bands are that they are wide bandwidths because they need to cover enough to get the chlorophyll content of leaves or something like that. Instead of getting the information that is present in the electromagnetic spectrum it would differentiate between different vegetation types, it sort of globs it all together. It averages all the radiance in broadband. Therefore, the difference between hyperspectral and traditional remote sensing is that instead of broadband where we have one number that represents chlorophyll, we will have a full spectrum that will allow us to differentiate for example, between different types of plants. This also helps in differentiating between different rock types.

It can be mentioned that the Ultra Violet atmospheric absorption is too great. There is limitation to receive a big information especially for a space platform. Therefore, it is generally blue through the near infrared, up to about two μm.

With the hyperspectral bands, the new high-resolution sensors, and what we call a virtual 3-D Geographic Information Systems, people can immerse themselves in a virtual reality geographic data system. A Geographic Information Systems and remote sensing data are going to be more understandable because it will look like as when we look out our window.

5.6. The Use of Remote Sensing and Geographic Information Systems in Construction Process

Remote Sensing method is the convenient source of appropriate data access in the wide area of applications. It opens an opportunity to embrace data needs. An accuracy of the method depends on the referenced data sources and quantity of applied data. At the same time, Geographic Information Systems developed by approaching to the data collected on the base of remote sensing method are one of the fastest growing computer-based technologies of past two decades; yet, full potential of this technology in construction has not been realized.

Based upon Geographic Information Systems capabilities, construction site layout is one of the areas that Geographic Information Systems could be applied successfully in all the stages of engineering activities. The layout of temporary facilities (TFs) such as warehouses, fabrication shops, maintenance shops, concrete batch plants, construction equipment, and residence facilities has an important impact on the cost savings and efficiency of construction operations, especially for large projects [110]. The primary objectives of this approach are to describe Geographic Information Systems technology and to present application of Geographic Information Systems technology to construction site layout. An example application of Geographic Information Systems to location optimization of any facilities in particular tower crane and concrete batch plant is provided to demonstrate Geographic Information Systems capabilities of use such a technology in the area, which are important part of construction management process as a whole. The spatial and no spatial data, which used in construction site layout process, demonstrates advantage and usefulness of Geographic Information Systems technology and environment and results shows that the Geographic Information Systems can solve site layout problems in a wide aspect of engineering services.

The data accessed by Remote Sensing method with further integrated into Geographic Information Systems technology is being used in diverse application areas related civil engineering, such as: disaster management systems, environmental problems, hydrologic, Infrastructure management systems, irrigation and drainage, transportation and urban planning. Directions magazine estimated that there were over one $ billion in Geographic Information Systems software sales alone in 2001, with additional billion spent for services, hardware, and related activities [111]. Although Geographic Information Systems technology have more capabilities on construction, such as network functions that assists in scheduling methods like CPM and PERT of which full potential has not been realized.

There is data of Remote Sensing and Geographic Information Systems technology application [112] used in construction stages for contractor prequalification. At the same time, some other authors [113] developed an integrated Geographic Information Systems that aids managers in activities related to onsite soil condition. Cheng and Yang presented the system, Material Plan, which integrated Geographic Information Systems - based on cost estimates with construction scheduling to automate planning tasks required for materials layout [114]. It has been investigated the roles of internet-based Geographic Information Systems in e-commence systems and used of Geographic Information Systems capability to integrate, to manage, and to analyse spatial information as well as has been considered the integration of Geographic Information Systems with visualization techniques for planning and visualizing of dam construction processes [115-116].

Furthermore, Poku and Arditi has developed a system called PMS - Geographic Information Systems (Progress Monitoring System with Geographical Information Systems) to represent graphical construction progress [117]. It is important to indicate that an automated system has been developed for construction temporary facilities. The system, ArcSite, including a Geographic Information Systems integrated with database management system (DBMSs) as the necessary stage of data processing. An ArcSite is a useful tool for construction site layout but it has not been used properly Geographic Information Systems capabilities. In addition, the location of construction equipment, as an important factor, is not considered in the system and the resulting layout of facilities may not arrive at the optimal solution. It is evident that with passing of about fifteen year from presenting the system, changes are required. A Geographic Information Systems have not been used, however, to solve the site layout problem as defined in this paper [118-119].

The primary objectives of this chapter are:

(1) to present a brief description of Geographic Information Systems technology;
(2) to describe the circumstances (methods) of carrying out the layout of construction temporary facilities;
(3) to explore potential Geographic Information Systems application to the construction site layout;
(4) to identify areas of additional research.

5.6.1. Description of Geographic Information Systems Technology

Among different descriptions presented about Remote Sensing and Geographic Information Systems technology applications where defined proper one as " Geographic Information Systems are a system of hardware, software, people, organization and institutional arrangement for collecting, storing, analysing, and disseminating information about areas of the Earth" [120].

To work with Geographic Information Systems, like other systems, familiar with database structure is required. Three basic types of database structures which more common are:

- hierarchical data structures;
- network systems; and
- Relational database structures.

Data in hierarchical data structures have one-to-many or parent-child relationship. In the hierarchical system, each relationship be explicitly defined before the structure and its decision rules are developed.

Network systems are an improvement over hierarchical structures for Geographic Information Systems work because they are less rigid and can handle many-to-many relationships. This structure allows users to move from data item to data item through a series of pointers. Many Geographic Information Systems databases have many-to-many relationship, for this reason network systems recognized as a proper structure.

Relational database structures, on the other hand, obviated the network systems shortcoming to used large numbers of pointers.

The database structure stored data as ordered records or rows of attribute values called tuples. The data included on life cycle of a facility can be categorized as spatial and no spatial. As much as 75% of the spatial data pertaining to a facility are common to all phases of the facilities delivery process [121].

A Geographic Information Systems, on the other hand, provide capabilities for spatial and no spatial data to be collected from different sources, stored, analysed, and presented systematically. Spatial data have four basic components that significantly differentiate them from any type of data:

- Geographical positions;
- descriptive attributes;
- spatial relationships, i.e., topography; and
- temporal relationships.

A Geographic Information Systems recognize three primitive types of spatial objects:

- points;
- lines; and
- areas.

Vector and raster data models are two alternate models for representing these objects. The basic logical units of the vector model are points, lines, and polygons. A point is stored as a single X, Y coordinate pair, a line as a series of X, Y coordinates, and a polygon (area) as a closed loop of X, Y-coordinate pairs defining the boundaries of the area. The basic units of the raster model are individual cells. The raster space is subdivided into cells (usually square). A row and column number references each cell. Additional attribute information concerning a given cell is stored with the data record for that cell. In this model, a point is represented by a single grid cell, a line by a number of consecutively neighbouring cells, and an area by a cluster of neighbouring cells.

The selection of a data model, vector or raster, depends largely on the selected construction application. Advances in hardware and software technologies, however, have gradually made the differences between them less significant. Raster model has advantages such as simplicity, low cost, ease of data collection, and image interpretive aspect, and for vector model could be pointed some advantages such as high accuracy and topological capabilities. To extract meaningful information from the Geographic Information Systems database, appropriate functions must be performed. This process is commonly referred to as data manipulation and analysis.

According to condition of problem, various types of Geographic Information Systems analysis functions can be performed. Some of the most important functions of Geographic Information Systems include:

- overlay operations;
- neighbourhood functions; and
- Connectivity functions, such as proximity and network operations.

Overlay operations involves the integration of multiple layers to create a new data set. The overlay functions can perform simple operations such as laying a road map over a map of local wetland. Resultant layer contains new spatial objects with different attribute and topological information.

Neighbourhood function analyses the relationship between an object and similar surrounding object. This function idea is to characterize each object as a part of a larger neighbourhood of objects based on some shared attributes. All neighbourhood processes require three parameters:

(i) the location of one or more target areas;
(ii) the size of the neighbourhood surrounding the target area; and
(iii) a function to perform in the defined neighbourhood and assigned to the targeted area.

The main role of proximity functions is to determine if a given point, line, or area lies within a certain distance of another point, line, or polygon. Network functions, on the other hand, can be used to allocate a given resource, such as construction material, or to determine the minimum cost path through a given network between an origin and destination. They are commonly used for moving resource from one location to another. Network analysis entails four components:

(i) set of resources;
(ii) one or more locations where the resources are located;
(iii) an objective to deliver the resources to a set of destinations; and
(iv) A set of constraints that places limits on how the objective can be met.

Data output is the procedure by which Geographic Information Systems results are displayed in a form suitable to the user. The most common output products generated by Geographic Information Systems are maps of various kinds, such as thematic maps, choropleth maps, isorithmic maps, dot maps, line maps, animated maps, landform maps, and cartograms. For applications such as engineering, other non-map graphics may be preferred. A Geographic Information Systems have the capability to generate several graphic formats. These graphics include - bar charts, pie charts, scatter plots, and histograms.

5.6.2. Construction Site Layout

Site layout planning is an important task that involves identifying the temporary facilities (TFs) needed to support construction operations, determining their size and shape, and appropriately positioning them within the boundaries of a construction site [122]. Temporary facilities range from simple lay-down areas to warehouses, fabrication shops, maintenance shops, concrete batch plants, construction equipment, and residence facilities [123].

Various approaches have been described in the literature to deal with the facility layout problem from different perspectives. Heuristic, solutions as opposed to optimum refer to solutions that are based on heuristic rules, experience, and simplified approaches. Due to the complexity of the construction site layout problem, mathematical optimization was only successful for a single or very limited number of facilities.

The majority of researchers, however heuristic approaches and knowledge-based systems have used to place the facilities on a construction site in a loosely packed manner. A loosely packed arrangement of facilities, as opposed to a tightly packed one, refers to the situation when spaces are permitted between facilities. Loosely packed arrangements are most suitable for laying out construction sites [124].

According to the methods of location temporary facilities, writers considered these facilities as three parts; tower cranes, concrete batch plant, and temporary buildings (e.g., office, parking, storage, shops etc.). In the next sections each of parts are described.

5.6.3. Construction Facility Layout Development

Tower cranes are the most expensive and frequently shared resources on building construction sites. Efficient utilization of tower cranes greatly depends on type, number, and location of tower crane(s). As the number of work tasks and the demand for tower cranes increase, planners may be required to make bold decisions on job conditions for a particular situation. A poor decision is likely to have significant negative effects, which will lead to additional costs and possible delays [125].

On large construction projects, several cranes generally undertake transportation tasks, particularly when a single crane cannot provide overall coverage of all demand and supply points, and/or when its capacity is exceeded by the needs of a tight construction schedule. Many factors influence tower crane location. In the interests of safety and efficient operation, cranes should be located as far apart as possible to avoid interference and collisions, on the condition that all planned tasks can be performed. However, this ideal situation is often difficult to achieve in practice; constrained workspace and limitations of crane capacity make it inevitable that crane areas overlap. Subsequently, interference and collisions can occur even if crane jibs work at different levels [126].

5.6.4. Concrete Batch Plants Development

A concrete batch plant is an important element in any concrete construction process, whether it is working as a central mixing plant onsite or offsite, supplying ready mixed concrete to a project. The concrete can be mixed in a batch plant offsite or in a batch plant onsite and then transported by transit mixers [127]. The major decision variables that influence management decisions of the concrete batch plant location are the volume of concrete consumption and the time of transportation between concrete batch plant and demand place.

Demand places are permanent facilities, which designed before construction begins. Therefore, their required concrete and their location have been defined. The objective function for batch plant location optimization was defined as minimum time-amount of concrete delivery. The shorter the time-amount of concrete delivery, the smaller the concrete transportation cost. Since concrete carrying machines (i.e., transit mixers) are similar in individual project, "time" replaces with "distance." Application of distance as criterion of objective function increased possibility of using more models.

The larger the distance of concrete delivery is the greater the transportation time and vice versa. Similarity of concrete carrying machines concludes linear relation between "time" and "distance."

One simple model for determining optimum location for concrete batch plant in site, minimizes the product of amount by distance. Between locations, which could use for batch plant location, the model should find one that has minimum amount of required concrete multiplied by distance.

5.6.5. Process of Temporary Buildings

This part of temporary facilities involved warehouses, residence facilities, office, parking, fabrication shops, maintenance shops, and labs. Finding optimizes location for temporary buildings depends on the facilities that temporary buildings supported them. Because of increasing productivity, minimizing travel time and improving safety some relations between permanent facilities and their support facilities have been developed. We could categorize these relations as "close to." "far from," "within specified distance of," "next to" and "visible from." Most TBs needed to "close to" their supported PFs. For example, the warehouse should be located close to entrances to ease deliveries [128]. In practice, material storage yards and warehouses are located as close as possible to the work areas to reduce excess on-site transportation. "Far from," as opposed to close to, used in situation that something endanger TBs. For example, OSHA standard 1926.407 recommends storage facilities of electrical equipment and possible sources of sparks are located far from flammable material [129]. Similar condition exists for hazardous material. These materials include:

- explosives and blasting devices used in rock excavation;
- flammable material such as fuel used by construction equipment;
- toxic substances such as asbestos, coal tar pitch volatiles, cadmium, benzene, formaldehyde, methyl chloride among other materials including 13 carcinogens identified by OSHA standards.

It is better to locate site offices and residence facilities outside the range of crane operations whenever possible [130].

Closeness sometimes represent quantitative. In this situation, one facility should be located "within specified distance of" another. The best example of the former state is for tower crane and supply point. The use of supply point within operating radius of tower crane is necessary. Some facilities have joint operations. These facilities should be located next to each other, like steel storage and wedding shop. For maintaining security, finally, some facilities should be located visible from another, like warehouse from office. Very likely satisfaction of one relation, invalidate another. For this reason planner should make priority. After making priority and sometimes use affinity or proximity weights by the planner(s), different systems could be applied for solving site layout problem. Several approaches have been developed in the literature to deal with this problem by different AI techniques-such as knowledge-based systems, [131-141] neural networks and genetic algorithms that have been explored in the development of site layout systems. In another research, it has been established assignment of priorities based on a satisfaction rate and weight of specified facility defined for another [128].

5.7. Construction Site Layout Development with Remote Sensing and Geographic Information Systems Application

Inability to apply different shapes for site and facilities is one of the problems in previous models to construction site layout. In most efforts only a rectangular shape or group of unit areas, which take a rectangular shape, are used. Generally, "closeness" criterion exists on the site layout models. A Geographic Information Systems models, also, are not exception of this principle. According to type(s) and application(s) of facilities, the closeness can be determined by using quantitative or qualitative methods. Quantitative methods may consider the actual transportation cost per unit time or the amount of materials moved per unit time between shipping and demand areas on site.

5.7.1. Approach of Development

Alternatively, qualitative methods consider a subjective numerical proximity weight to express the desirability of having any two facilities close to each other on the site layout. Both methods could be used in the Geographic Information Systems models. Location optimization process for tower crane(s) mainly use spatial data. Feasible task area, in first, determines based on location of demand and supply points and crane's operating radius. These are spatial data. Proximity functions determine area that lies within a certain distance of supply and demand points according to cranes type.

Neighbourhood function applied to grouping tasks. After determining initial tasks, Geographic Information Systems model assigns those respectively to different groups as the first member of the group, and then other tasks according to overlapping area with each group cluster to them by neighbourhood function. This process requires location of initial tasks and the size of overlapping area, which is determined by proximity functions. Finally, overlay operations could be used to eliminate shortcoming of previous models. It was necessary to predetermine the type and geometric layout of all supply points in past models. Sometimes applied different types of cranes resulted in reduction of cost or increase efficiency. In addition, supply point may change during construction phase. With aiding of overlay operation in Geographic Information Systems model, possibility of existence of different crane types and supply point are prepared. Each type of crane (or supply point) has placed into different layer and with network systems this process could be made.

Concrete batch plant is one of the temporary facilities showing potential application of Geographic Information Systems in site layout efficiency. As noted before, minimize transportation cost is a major factor in batch plant location optimization. Therefore, the objective function was defined as minimum time-amount of concrete delivery. When time replaced with distance, GIS application was possible. Some similar applications of Geographic Information Systems have been addressed in the previous researches.

Locating fire stations, schools and other facilities according to the shortest distance (or time) between them and supported place is a common application of Geographic Information Systems. Network functions can be used to allocate a concrete batch plant based on determining the minimum cost path through site roads between batch plant and demand points, i.e., permanent facilities. For this purpose, one or more locations where the batch plant could be located must be determined. Also set of constraints that places limits on how the concrete can be met and the objective function to deliver the concrete should be determined. A Geographic Information Systems aided with amount of concrete required by permanent facilities, presented location which minimum distance multiple by amount of required concrete from demand points. This location refers as an optimum location of concrete batch plant.

5.7.2. Areas for Future Research

Based on existing knowledge and understanding of Geographic Information Systems technology, several research areas about construction site layout can be identified. First, for proving Geographic Information Systems capability on tower crane locating, described procedure should be done by Geographic Information Systems technology. In addition to compare, the Geographic Information Systems result with the results of previous models, different types of crane and different layout of supply points added to Geographic Information Systems model. This act results in eliminating previous models shortcoming. For the reason that location optimization of concrete batch plant is different procedure and the use of another Geographic Information Systems analysis function, the next model allocated to this temporary facility. This model is not new to Geographic Information Systems professional users but illustrates Geographic Information Systems capability efficiency.

Another model was assigned to layout of temporary buildings. Similarities between Geographic Information Systems and other computer systems, such as computer-aided design (CAD), project management (PM) systems, and facility management (FM) systems, the Geographic Information Systems could be applied in planning of construction site layout. Relation requirements between temporary buildings and their supported facilities, have an excessive impact on results.

Finally, combining these models to a comprehensive model, implementation of all the layout process becomes possible. Thus, Geographic Information Systems is the only system, which could be applied for locating whole of temporary facilities such as tower crane, concrete batch plant, and temporary buildings.

This briefly describes Geographic Information Systems technology and its analyses functions. Data required for layout of temporary facilities (TFs) has been gathered and according to the methods of location, temporary facilities categorized into three parts; tower cranes, concrete batch plant, and temporary buildings (e.g., office, parking, storage, shops and etc.). The manner of locating for each part was described. Then potential applications of Geographic Information Systems technology in construction site layout were identified. Specific emphasis was placed on the spatial aspects of these applications as the justification for the use of the technology. Selected applications, location optimization of tower crane, concrete batch plant, and temporary buildings, were used to highlight the strength of the technology. Suitable analysis functions, such as overlay operation, proximity and neighbourhood function were suggested for each category of temporary facilities. Writers attempted to show coordination of Geographic Information Systems technology with data required in construction site layout. For this reason, a section of the paper was allocated for exploring the potential of the Geographic Information Systems application to the construction site layout. Finally, areas of additional research were identified.

5.8. Remote Sensing and Geographic Information System Applications in Industry

Geographic Information System is a computer system for capturing, storing, quarrying, analysing, and displaying geographic data. Like any information technology, Geographic Information Systems is a special class of information system. A Geographic Information Systems activity can be grouped into spatial and attribute data management, data display, data exploration, data analysis and Geographic Information Systems modelling (Lo and Yeung, 2002) [142]. A Geographic Information Systems involve in using both spatial and attribute data, spatial data relate to the geometry of the features while attribute data describes the characteristics of the space features. Attribute data are stored in the tabular form, where each row of a table represents a feature while column represents the characteristic of features.

The intersection of a column and a row shows the value of particular characteristics of a feature. Two (spatial and attribute) sets of data files are synchronized so that both can be quarried, analysed and displayed [143]. A Geographic Information Systems is used in various disciplines but its application is new to the field of construction industry. Literature suggests a limited use of Remote Sensing method and Geographic Information Systems technology in this discipline. The reflection of the following aspects have been considered:

- discusses about the various applications of Geographic Information Systems in construction industry;
- a methodology to create 3-D view of a sample building;
- An approach to represent a construction schedule in Geographic Information Systems environment is discussed which is followed by a summary of conclusions.

5.8.1. Remote Sensing and Geographic Information Systems in Construction Industry

In construction industry the information required for planning and design are stored in different form, such as drawing, specifications, and bar charts. In planning process, the planner has to repetitively reorganize and interpret the information collected from various resources. This process is tedious and prone to errors [144]. Thus, construction industry require a system which should be capable of integrating various types of data and provide the required information and data timely that will finally support various decision and construction operations. With the advances in the field of information technologies, construction industry has started taking advantages of some of these developments. A Geographic Information Systems is a new tool in information technology and can improve the construction planning and design efficiency by integrating locational and thematic information in a single environment. The capabilities of Geographic Information Systems to store a large database can be utilized to keep construction data in digital form also [145]. The construction database can provide a wide range of information to the construction industry with a mechanism for rapid retrieval and manipulation capability. Use of Geographic Information Systems may also satisfy the need of spatial and descriptive information required in different construction process [146]. The prototype system developed industry [147] suggests that instant spatial data capture provides fast and accurate visual information about the progress on the construction site. Integration of spatial database with project management functions provides a powerful and effective management control system. It has been [148] suggested the use of database management capabilities of Geographic Information Systems to develop a 3-D subsurface environment from well-log data collected from a series of boreholes. Because of the surface and subsurface conditions influence the construction methods and choice of equipment to be used, which, in turn, affect the cost and scheduling of projects, it is important that site conditions should be properly assessed. A database for the storage of descriptive soil data in Geographic Information Systems environment have been used [149] in some of studies for relating soil data to display the corresponding locations of boreholes on the map.

5.8.2. Geographic Information System Based Approach of Evaluation

In the meantime, it was suggested Geographic Information Systems - based on the approach for quantity take-offs and cost estimation. The method involves in dividing the architectural drawing into different layers, called data layers. An area and perimeter are used as the basic parameters for quantity take-offs. Thus, data layers are created as polygons in AutoCAD and transferred to Map/Info as geometric coverage. Locational information includes spatial features such as coordinate, area, perimeter and spatial relationship, which are derived from the coverage. Whereas thematic information includes identification code, beam number, floor number etc., this information is entered by the user. Locational and thematic information are integrated by using one-to-one, one-to-many, and many-to-many relationships and is achieved by an Open Database Connectivity (ODBC) technique. Spatial operations are performed to identify the required geometric dimensions of the graphical features. To complete the quantity calculation, the user inputs the parameters such as the depth of beam and slab, floor height, and area of door and windows. The Structured Query Language (SQL) is used to retrieve the data for quantity take-offs from the related attribute tables [143].

In temporary facilities layout the project manager has to continuously extract information from various resources and draw it on the paper [150]. As temporary facilities should be located as close to their supporting activities to reduce the time for travel, Cheng and O'Connor [150] developed an automated site layout system called ArcSite using Geographic Information Systems for construction of temporary facilities. ArcSite consists of Geographic Information Systems integrated with DBMS to identify suitable areas for the location of temporary facilities.

ArcSite integrate the information required for finding suitable location for temporary facilities and performing a series of complicated spatial operations and database queries to identify optimal site, which is quite difficult to perform manually.

Cheng and Chen [151] developed an automated schedule monitoring system by using Geographic Information Systems to assist the construction managers to control the erection process for precast building construction. A case study is taken where structural elements were prefabricated in the manufacturing plant and transported to job site for installation. The schedules for prefabrication and transport of the structural elements to the job site are developed based on installation schedule. The study suggested that the use of Geographic Information Systems environment improves the real time schedule monitoring system and construction process as well as improves the construction efficiency. The barcode system combined with wireless radio technology for identification of precast member is applied to collect and transmit the job site data to the control centre automatically. Through a real time monitoring of the construction process the scheduled components for erection are repetitively tracked and well controlled to assure that the lifting schedule is implemented as planned [151]. Study reported by Li, Chen, Yong, and Kong (2005) [152] integrates Global Position System and Geographic Information Systems technologies to reduce construction waste. A system was developed from automatic data capture system, barcode for construction material and onsite equipment management, while integrated Global Positioning System and Geographic Information Systems technology for material and equipment based on the Wide Area Network (WAN). Global Positioning System and Geographic Information Systems integrated with construction material and equipment management system are developed in such way that manager from headquarter and construction sites get real-time information to control cargoes on the road to the sites.

Comparison of the non-integrated system versus the Global Positioning System and Geographic Information Systems integrated system suggests that the Global Positioning System and Geographic Information Systems integrated solution improved the construction efficiency by increasing the effective working hour of construction equipment thus reducing the construction duration as well as the cost of workforce.

Li, Kong, Pang, Shi and Yu [153] presented an Internet-based Geographic Information Systems model for e-commerce business if link between buyers and suppliers is through electronic markets. The e-commerce system, called COME (construction materials exchange) was thus developed which can be used for on-line order and off-line delivery of different construction materials. The electronic market provides platform for the suppliers to provide online information about their products. Buyers can easily search and compare products of different suppliers through online system and contacts of the suppliers directly. If required, buyers can also invite agents to undertake certain tasks required in order to complete a transaction.

In all e-business activities, where transportation cost is involved, spatial information plays an essential role. Thus, internet-based Geographic Information Systems provide an ideal solution to manage costs of transportation and market analysis in the overall e-commercial activities [153]. The costs for transporting construction materials are not only dependent on the distance but it may involve many other variables, such as the locations of local distributors, reduced shipping costs because of combining shipment to various buyers in the same area. The author suggested further improvement, which could be carried out in the proposed system (COME) by taking care of these problems.

Varghese and O'Connor [154] developed a system with a platform on which the information required for route planning can be integrated. They successfully demonstrated the value of expert of Geographic Information Systems for automating the tedious and respective route planning tasks for large vehicles within the construction site. The study by Cheng and Chang [155] discusses the development of Geographic Information Systems based on system to automate the process of routing and design of an underground power supply system. In urban areas, obstacles such as existing public utility lines, railways, canals, and roads can influence the route significantly, as there may only be a limited number of feasible crossing points. Selection of a suitable route to avoid existing obstacles not only reduces the risk of damaging existing utilities, but also minimizes the cost and duration required for construction. In developed system, surface and underground utilities are represented in several coverages. The optimal paths for routing are determined using the network analysis of Arc/Info Geographic Information Systems package. The conflict points between the basic coverages and the selected route are identified and a reallocation schedule is determined through database queries and spatial operations.

The simulation process has proved to be an effective tool for planning and improving the performance of a construction process in many successful case studies. However, these tools lack the capability to represent explicit information involved in the simulated construction process. Zhong, Li, Zhu, and Song [156] suggested that Geographic Information Systems could be utilized to overcome these limitations. 3-D spatial data is used to represent elements that have physical dimensions in 3-D Geographic Information Systems. By considering time as an attribute of the 3-D spatial model, Geographic Information Systems can depict the simulated operations dynamically in a 3-D environment as being carried out in the same way, as they would be in the real world. This process of visualization can help to detect performance inconsistencies and obtain insight into the simulated construction operations.

5.8.3. Implementation of Processing

In spite of widespread applications of Geographic Information Systems to construction industry, visualization using Geographic Information Systems has not yet been used to its maximum potential in construction planning. Most of the available simulation tools are capable of optimizing construction sequences and establishing project plans. However, these tools are not having the capabilities to represent information involved in the simulated construction process.

Project managers are forced to search for design drawings and data reports that are needed while using simulation tools. So Geographic Information Systems can be utilized to overcome the above-mentioned limitations. Construction industry uses different tools other than the Geographic Information Systems for visualization, which are not capable to store the huge amount of information involved in any project. As the construction project duration increases, manager receives more spatial and non-spatial data, thus, manager has to take the decisions after careful investigation and analysis of a large amount of data.

Researchers have explored the capability of Geographic Information Systems in construction database management, and not much in visualization. A simple application presented here discusses the methodology to generate a 3-D view of building in Geographic Information Systems environment. Although the CAD technologies provide alternative for this, but Geographic Information Systems technology uses data (i.e. non-spatial) synchronized with the 3-D modelling which is not possible in CAD technologies. A Geographic Information Systems is a database centric tool, thus, handle spatial, and attribute data on a single platform.

The CAD layers organize the spatial data only to facilitate drafting process, while in Geographic Information Systems various spatial operations on graphics and non-spatial operation on the attribute data are possible. Present study provides a methodology to represent buildings in 3-D, in which the different information are attached to corresponding components of building [157-158].

5.8.4. Geographic Information System - Based Method of Development of 3-D Building Visualization

To develop a methodology for 3-D visualization of a building ArcView 3.2 is used and the procedure involves following steps:

- **Step 1:** As an architectural drawing it is not enough to generate a 3-D view in Geographic Information System, thus, the drawing is divided into different layers. The purpose of dividing of architectural drawing into different layers is to determine how layers will represent the complete 3-D view in Geographic Information System. All features of different layers are created in AutoCAD as polygon. Different layers such as base concrete, footings, walls, slab, doors, and windows include a number of polygons. Three wall layers are created at different height intervals so as take care of any opening in the wall. Figure 19 shows the different layers of sample building created in AutoCAD and used to generate 3-D view in Geographic Information System [159].
- **Step 2:** All the data layers are then transferred into ArcView. Figure 20 shows the different layers of the sample building transferred into Geographic Information System. Different polygons in a layer are dissolved to form a single polygon using dissolve of Geo-Processing extension of ArcView. Dissolve is repeated for all data layers. After dissolve, all AutoCAD files are converted as the shape files in ArcView. Shape files are a simple, non-topological format for storing the geometric location and attribute information of geographic features in ArcView. Vector data model is

involved in using both spatial and attribute data in a way that spatial data relates the geometry of the feature in layer, whereas attribute data describe the characteristics of layer's feature. There is always a dynamic linkage between the spatial and attribute data where highlighting a record of the attribute data highlight the corresponding spatial feature in view window of ArcView and vice- versa.

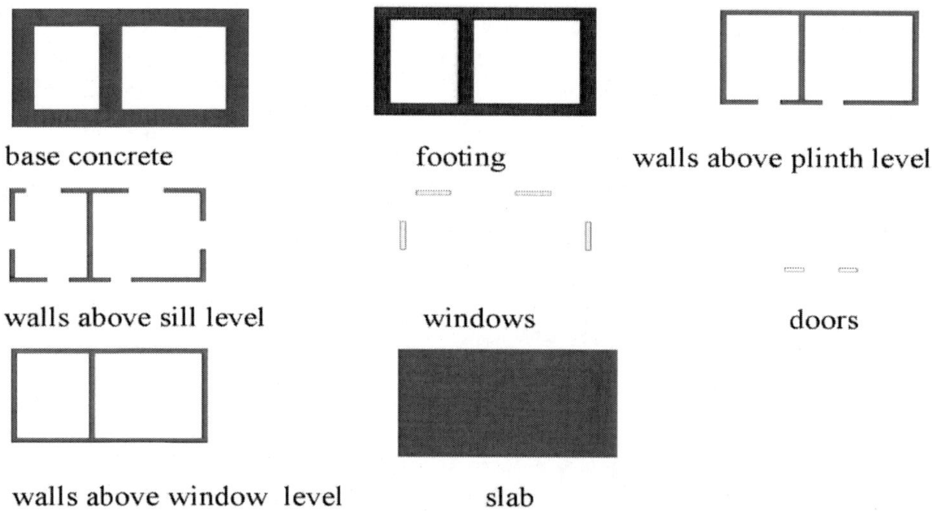

Figure 19. Different data layers of architectural drawing used to generate 3-D view.

- **Step 3:** The default sets of attributes are automatically generated when different layers retransferred into ArcView. Attributes needs to generate 3-D view included base-height and layer-height. Base-height is defined as the elevation value for a shape files. Thus, field called base is added into the attribute table of each layer that defines the base-height (elevation) and field height is also added into attribute table of each layer that gives the layer-height (i.e. thickness).
- **Step 4:** A 2-D layer does not have the base-height and layer-height information. To display the perspective view, a 2-D layer must be assigned base-height and layer-height from another source. The source will be the fields in its own attribute table. Thus, 2-D theme is assigned the values of base-height and layer-height from its own attribute table. A dialogue-box as shown in Figure 21(a) is utilized to assign the fields' base and height to the 2-D layers. Extrusion utility in the ArcView changes the form of a feature from points to vertical lines, lines into vertical walls and polygons into 3D blocks. Figure 21(b) shows different layers in space at an elevation equal to their base-heights value and extruded upward to construct a 3-D view, by a value equal to its layer-height. Figure 22 provides different stages of construction process for the sample building, whereas Figure 23 shows the 3-D view of the sample building.

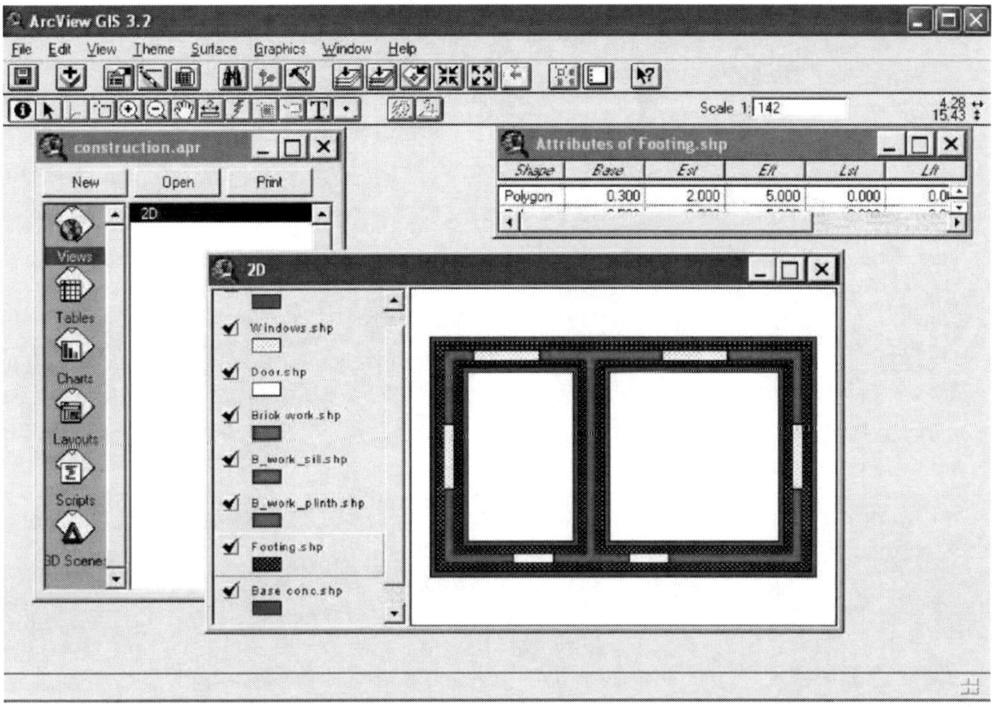

Figure 20. Spatial and non-spatial information stored as 2-D layers in ArcView, which is used to generate a 3-D view in Geographic Information System environment.

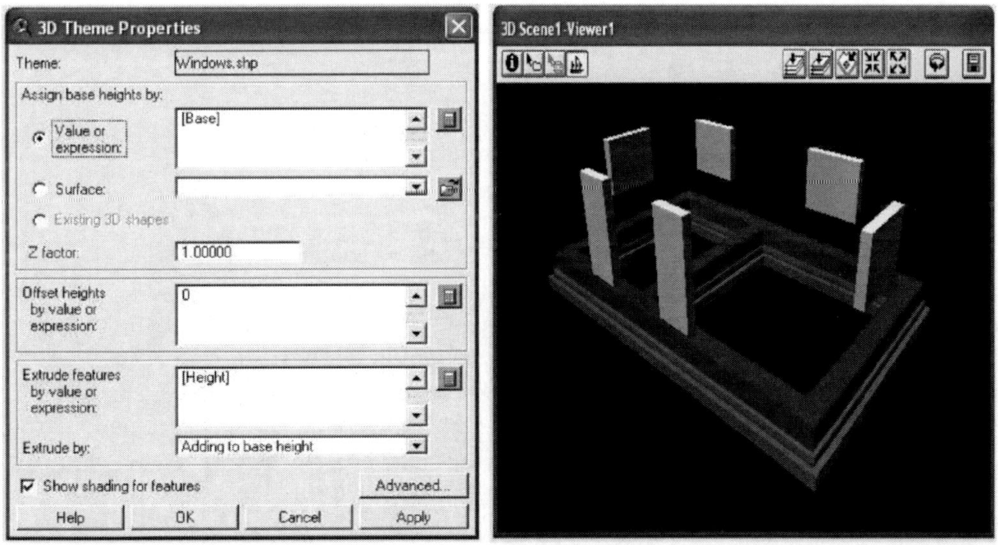

Figure 21. (a) 3-D properties dialog-box in ArcView used to assign the base-height and layer-height and (b) different layers in space at elevation value equals to its base height and extruded upward by value equal to its layer-height.

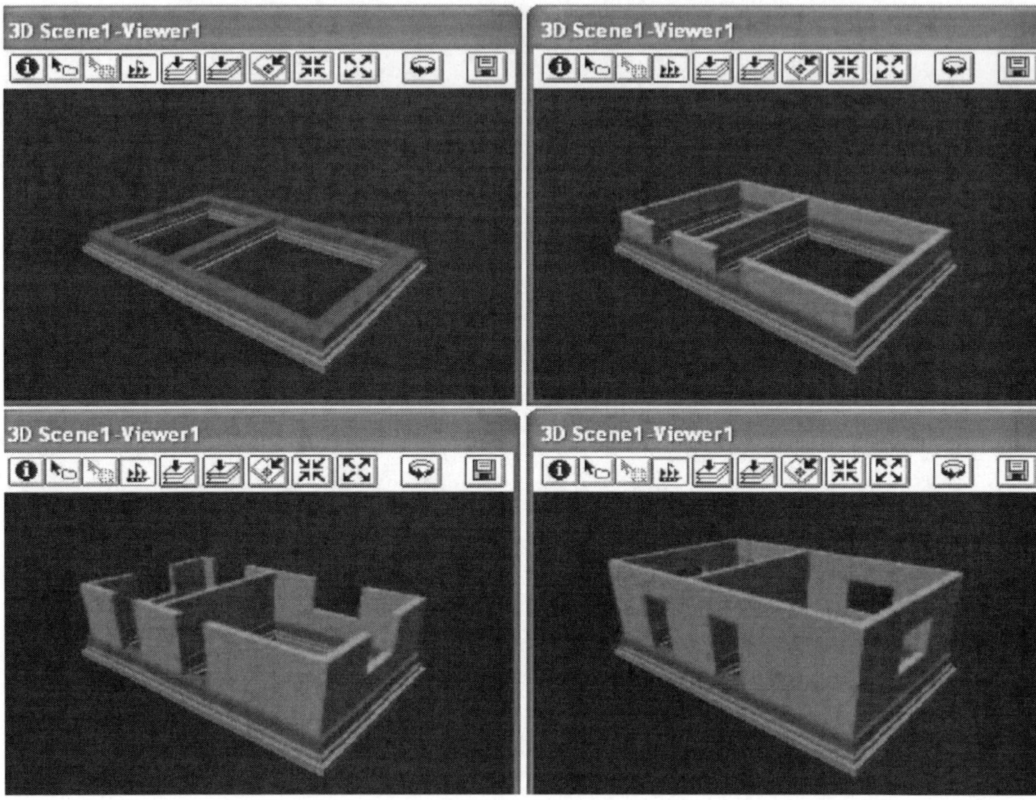

Figure 22. Appearance of building at different stages of construction using three-dimensional analysis of ArcView.

5.8.5. Geographic Information System Based Time Management Development

Schedule is defined as the work program or the timetable for the actions to be taken during the implementation of a construction project. Bar-chart method is one of the popular methods used by the contractors for scheduling. In bar chart method, work is firstly, splinted into different activities. These activities are then listed in the order of construction priorities on the left hand side column, while the time scale is plotted horizontally on the bottom. ArcView can be used to show the schedule using its in-built chart document. Figure 23 shows the ArcView's chart document, utilized to show the construction schedule of different construction activities.

The main advantage of the ArcView's chart document is when a bar on bar chart is clicked a window appears which provide the scheduling information related to that particular activity. Thus, with a single click on bar user can obtain the required information from the database [160-161].

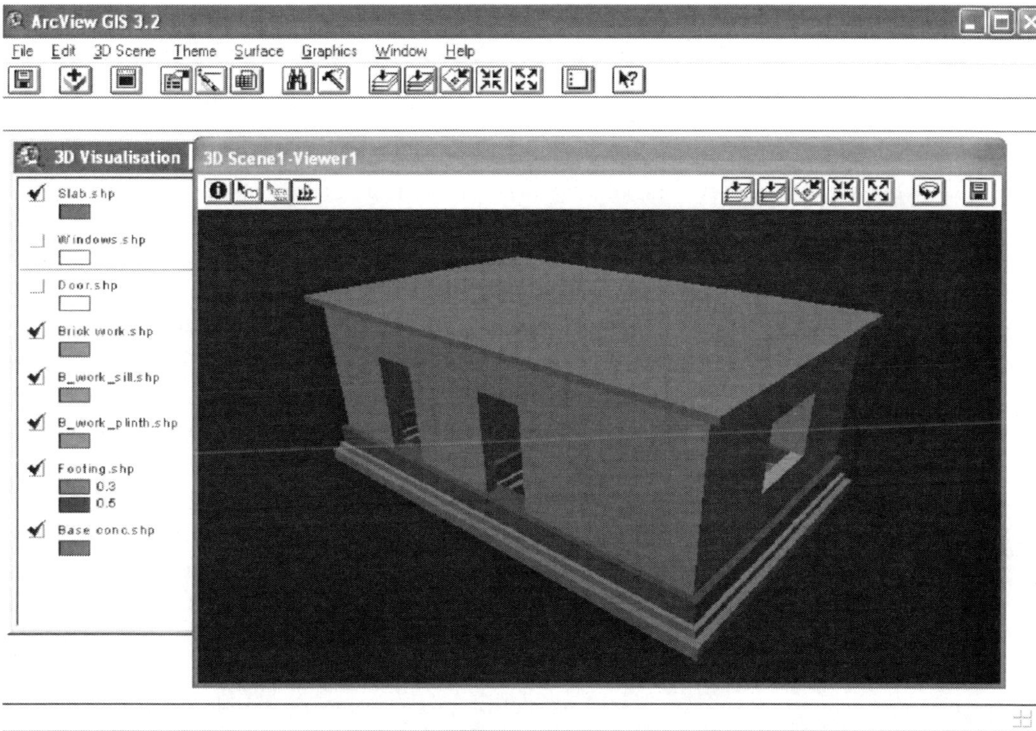

Figure 23. 3-D appearance of building prior to construction using 3DAnalyst of ArcView, relevant information is stored in the attribute table.

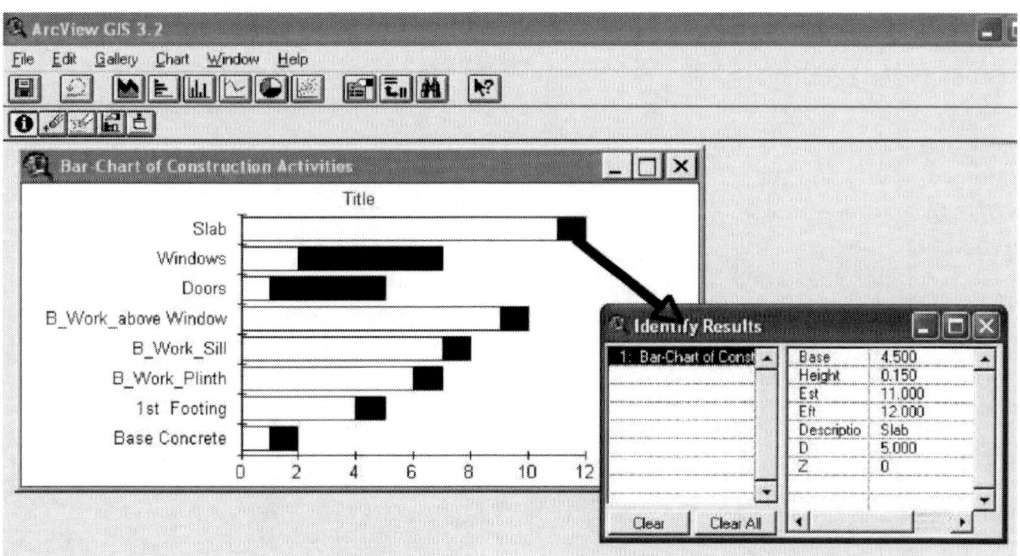

Figure 24. The Use of ArcView's Chart Document to represent the schedule of the different activities in sample application.

This described approach presents a survey of different applications of Geographic Information System to construction industry as well as methodology to generate 3-D view and bar chart to represent the construction schedule. A Geographic Information System is one of the fast emerging fields utilized in various engineering projects whereas its complete potential to the construction industry has not been realized yet. A Geographic Information System provides several benefits to the construction industry, in which most of the construction management software lack. A Geographic Information System may improve the construction planning and design efficiency by the integration of spatial and attribute information in single environment. A tool can effectively be used for quantity takeoffs, cost estimates, and database management. By integrating schedule with material requirement, dynamic material requirements plan can be developed in Geographic Information System environment. Database for soil data, foundation analysis and design, construction planning and design-construction integration may be developed in Geographic Information System environment.

Database in Geographic Information System environment can provide a wide range of information to construction industry with a mechanism for rapid retrieval and manipulation capabilities. Integration of schedule and design information makes it easier for the project manager to monitor and control the construction progress. Several tools for construction industry using Geographic Information System are suggested in literature and their applicability has been demonstrated with suitable case study. However, the practical usefulness of these developed tools in construction industry is still doubtful. Tool developed in one study is not utilized in the other study to develop another application and the tools implementations on the real life project in the industry are rare. Further, most of the reported works have used different software in combination with Geographic Information System software to develop different tools for construction industry, thus require studies to develop tools that involve in using new inbuilt scripts/codes in thc Geographic Information System environments itself to replace the use of other software.

The sample application presented here explains the Geographic Information System based procedure for 3-D visualization by utilizing the capability of Geographic Information System to maintain data in different layers. Data is well stored in the different layers in Geographic Information System and features in different layers are constructed as polygons. So these data layers can further be used for quantity take-off. Thus, Geographic Information System can also be used as an alternative visualization tool for construction industry. Although the CAD technologies provide visualization capabilities, but Geographic Information System technologies can perform different operation on the attribute data synchronized with the 3-D model, which is not yet possible in CAD technologies.

CONCLUSION

There is no doubt that Remote Sensing with further Geographic Information System development is an excellent opportunity in civil planning. It is a tool consisting of a large technological advances to integrate a big size of information, to manage and access when required.

Chapter 6

REMOTE SENSING AND GEOGRAPHIC INFORMATION SYSTEM METHODS IN TOWN PLANNING

ABSTRACT

Remote Sensing method and Geographic Information System technologies achievements have been developed in the late of 1960s, yet in the early days very few planning departments installed them because of the prohibitive cost of hardware and the limited capabilities of the software [162]. Most of the early software systems focused on the computer mapping with few analytical functions. The most powerful software at that time was grid based [163]. The subsequent fall in the prices of hardware, computer storage, and peripherals, accompanying improvement in the performance of hardware and software (particularly the speed of computer processors), and advances in the data structures and related algorithms of vector-based GIS [164], has made Geographic Information System more affordable, less time consuming and more workable. Since the early 1980s, there has been a marked increase in the installation of Geographic Information System in different levels and departments of urban and regional governments in the developed countries, notably of Europe and North America [165-167]: see Campbell, Chapter 44. With the further decrease in the price of computer hardware and software, the use of GIS has emerged in urban planning in the developing countries in the 1990s [168]. A Geographic Information System is increasingly accessible to planners and is now an important tool for urban planning in developed and developing countries alike.

Keywords: town planning, Geographic Information System, integration space data and field measurements, conceptual approach

INTRODUCTION

Honestly, it is High Technology application in town planning. Remote Sensing Geographic Information System is helpful in such a complicated area as town planning. Today existing requirements and regulations demand high standards in town planning. This limits the use of traditional technology and technical facilities in order to satisfy of desires and expectations.

In current stage, it is necessary to find out a new technology to meet modern requirements. At the same time, it does not limit only engineering requirements as well as opens an opportunity for a successful management of all the segments of engineering containing of town planning activities.

In this case, space technology applications based on Remote Sensing and Geographic Information System is capturing mainly modern demands in point of technological and management view.

6.1. Approach of Remote Sensing and Geographic Information System in Town Planning

A Geographic Information System is just one of the formalized computer-based information systems capable of integrating data from various sources to provide the information necessary for effective decision-making in town planning [169]. Other information systems for urban planning include database management systems (DBMS), decision support systems (DSS), and expert systems. A Geographic Information System serves both as a database and as a toolbox for urban planning. In a database-oriented Geographic Information System, spatial and textual data can be stored and linked using the geo-relational model. Current Geographic Information System supports efficient data retrieval, query, and mapping. Planners can also extract data from their databases and input them to other modelling and spatial analysis programs. When combined with data from other tabular databases or specially conducted surveys, geographical information can be used to make effective planning decisions. As a toolbox, Geographic Information System allows planners to perform spatial analysis using geo-processing functions such as map overlay, connectivity measurement, and buffering [170, 171]. Among the geo-processing functions, map overlay is probably the most useful tool.

6.1.1. Concepts of Use of Remote Sensing and GIS in Town Planning

Town planning is one of the main applications of Geographic Information System. Urban planners use Geographic Information System both as a spatial database and as an analysis and modelling tool. The applications of Geographic Information System vary according to the different stages, levels, sectors, and functions of urban planning. With the increase in user-friendliness and functions of Geographic Information System software and the marked decrease in the prices of Geographic Information System hardware, Geographic Information System is an operational and affordable information system for planning. It is increasingly becoming an important component of planning support systems.

Recent advances in the integration of Geographic Information System with planning models, visualization, and the Internet will make Geographic Information System more useful to urban planning. The main constraints in the use of Geographic Information System in town planning today are not technical issues, but the availability of data, organizational change, and staffing. Land suitability analysis, which is itself, an important component in urban planning [172-174].

Database management, visualization, spatial analysis, and spatial modelling are the main uses of Geographic Information System in urban planning. A Geographic Information System is used for the storage of land use maps and plans, socioeconomic data, environmental data, and planning applications. Planners can extract useful information from the database through spatial query. Mapping provides the most powerful visualization tools in Geographic Information System. It can be used to explore the distribution of socioeconomic and environmental data, and display the results of spatial analysis and modelling exercises. Spatial analysis and modelling are used for spatial statistical analysis, site selection, and identification of planning action areas, land suitability analysis, land use transport modelling, and impact assessment.

Interpolation, map overlay, buffering, and connectivity measurement are the most frequently used Geographic Information System functions in spatial analysis and modelling. The use of the above functions varies according to different tasks and stages of urban planning.

The many benefits in using Geographic Information System in town planning include [177]:

- improved mapping – better access to maps, improved map currency, more effective thematic mapping, and reduced storage cost;
- greater efficiency in retrieval of information;
- faster and more extensive access to the types of geographical information important to planning and the ability to explore a wider range of 'what if' scenarios;
- improved analysis;
- better communication to the public and staff;
- Improved quality of services, for example speedier access to information for planning application processing.

6.2. Remote Sensing Method and Geographic Information System Technologies Applications in Town Planning

6.2.1. Town Planning Process Segments Management

Urban planning involves many functions, scales, sectors, and stages. In general, the functions of urban planning can be classified into general administration, development control, plan making, and strategic planning. General administration and development control are relative routine planning activities, whereas plan making and non-routine strategic planning are undertaken much less frequently. The scale of the planning area covered can range from a whole city, to a sub-region of a city, a district, or a street block. The most frequently involved sectors of urban planning are land use, transport, housing, land development, and environment. At each scale of planning there are different stages: the determination of planning objectives; the analysis of existing situations modelling and projection; development of planning options; selection of planning options; plan implementation; and plan evaluation, monitoring, and feedback. Different functions, scales, sectors, and stages of urban planning make different uses of Geographic Information System.

The use of the data management, visualization, spatial analysis, and modelling components of Geographic Information System varies according to different functions of urban planning. Data management, visualization, and spatial analysis are used more in the routine work of urban planning. Spatial modelling is used more in a strategic planning. General administration employs mainly data management and visualization. Finally, development control uses the visualization and spatial analysis functions of Geographic Information System most. The routine general administration and development control work of urban planning includes [178]:

- management of land use records;
- thematic mapping;
- planning application processing;
- building control application processing;
- land use management;
- land availability and development monitoring;
- industrial, commercial, and retail floor space recording;
- recreational and countryside facility planning;
- environmental impact assessment;
- contaminated and derelict land registers;
- land use/transport strategic planning;
- public facilities and shops catchment area and accessibility analysis;
- Social area and deprivation analysis.

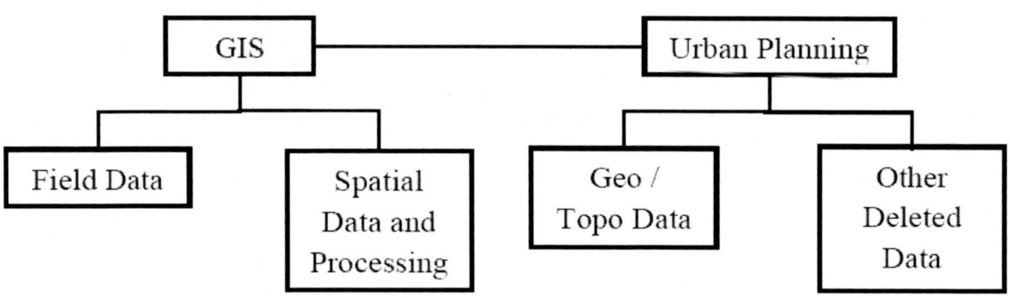

Figure 25. Layout of Geographic Information System use urban planning.

Visualization, spatial analysis, and spatial modelling are the most frequently used Geographic Information System functions in plan making. In the work of [164] the advantages of use of the data management, visualization, and spatial analysis and modelling functions of Geographic Information System as scientific inputs to urban planning are discussed. Webster shows that there are significant differences in the degree of Geographic Information System use in the description, prediction, and prescription planning process. Description is used more often in general administration, whereas prediction and prescription are used more often in plan making.

Different scales of planning require different data and techniques. Raster data are more useful for citywide strategic planning, because large areas are involved and high resolution is not required. The processing of raster data is much faster than that of vector data, especially in map overlay and buffer analysis. On the other hand, vector data are generally used for district and local action area planning because of the need for very high-resolution analysis.

There are many applications of Geographic Information System in the land use, transport, housing, land development, and environmental sectors. Key examples include site selection and land suitability analysis. In contrast, network analysis and route selection are most frequently used in transport planning, and environmental planning and management use buffer and overlay processing. There is an increasing trend toward the integration of modelling in different sectors of urban planning [179].

The role of Geographic Information System also varies in different stages of the urban planning process. For example, Geographic Information System is more useful in modelling and development of planning options than in the determination of planning objectives. The different stages in the urban planning process can be generalized as the determination of objectives, resource inventory, analysis of existing situations, modelling and projection, development of planning options, selection of planning options, plan implementation, and plan evaluation, monitoring, and feedback. A Geographic Information System can only provide some of the data and techniques that needed in different stages of the urban planning process. Any Geographic Information System also has to work with other databases, techniques, and models at different stages of the planning process.

6.2.2. Resource Management in Town Planning

Geographical information, when integrated with remote sensing, can save time in collecting land use and environmental information. Remote Sensing images are becoming an important source of spatial information for urban areas [180]. They can help to detect the land use and land use changes for urban areas [180]. In particular, stereoscopic pairs of digital aerial photographs can be used to derive 3-dimensional CAD models of buildings for dynamic visualization of a city, or for direct import into a Geographic Information System database [181].

6.2.3. Monitoring and Analysis of Condition

A Geographic Information System can help to store, manipulate, and analyse physical, social, and economic data of a city. Planners can then use the spatial query and mapping functions of Geographic Information System to analyse the existing situation in the city.

Through map overlay analysis, Geographic Information System can help to identify areas of conflict of land development with the environment by overlaying existing land development on land suitability maps. Areas of environmental sensitivity can be identified using Remote Sensing and other environmental information [182].

6.3. Future Planning

6.3.1. Forecasting of the Possible Changes in Town Planning

A key function of planning is the projection of future population and economic growth. A Geographic Information System can be used for prediction and projection [183]. Spatial modelling of spatial distributions makes it possible to estimate the widest range of impacts of existing trends of population, and of economic and environmental change. For example, a range of environmental scenarios can be investigated through the projection of future demand for land resources from population and economic activities, modelling of the spatial distribution of such demand, and then using Geographic Information System map overlay analysis to identify areas of conflict. Using socioeconomic and environmental data stored in Geographic Information System, environmental planning models have been developed to identify areas of environmental concern and development conflict [184]. A Geographic Information System can also be used to model different development scenarios. It can show the modelling results in graphic form, making them easy to communicate with the decision-makers [185, 186].

Planners can use such information to formulate different planning options and help to guide the future development so that they avoid such conflicts.

6.3.2. Stages of Town Planning

Land suitability maps are very useful in the development of planning options. They can be used to identify the solution space for future development [187]. The association of spatial optimization models with Geographic Information System can help to formulate and develop planning options, which try to maximize or minimize some objective functions [188]. The simulation of different scenarios of development with Geographic Information System can help in developing planning options [189].

6.3.3. Implementation of Town Planning Undertakings

The final selection of a planning option is increasingly a political process, but planners can provide technical inputs to this process in order to help the community in making their collective choices. The integration of spatial and non-spatial models within Geographic Information System can help to evaluate different planning scenarios [190]. The use of Geographic Information System with multi-criteria decision analysis can provide the technical inputs in the selection of planning options [191, 192].

Planning process contains the Geographic Information System, Remote Sensing and other databases and models including determination of objectives resource inventory analysis of existing situation, modelling and projection of development of planning options, selection of planning options, plan implementation/plan evaluation, monitoring/and feedback of Geographic Information System, Remote Sensing as well as other databases and models.

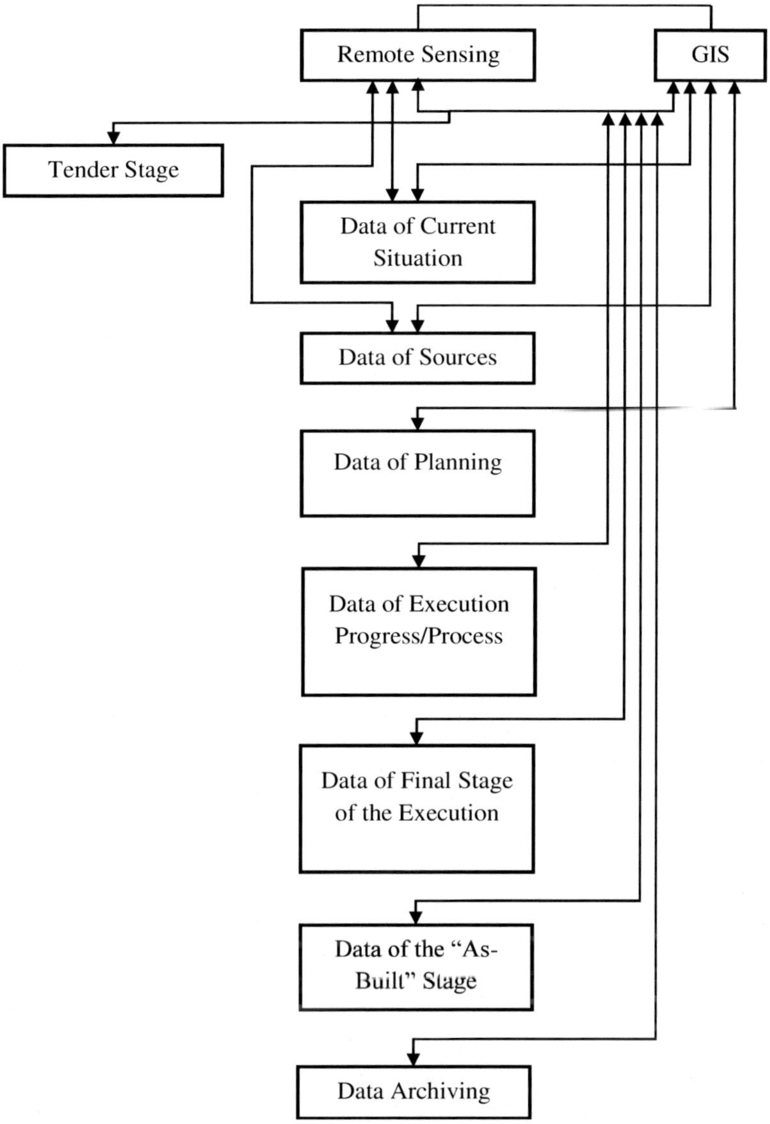

Figure 26. Remote sensing and Geographic Information System in planning process.

6.3.4. Monitoring and Evaluation Developments

A Geographic Information System can be used in the implementation of urban plans by carrying out environmental impact assessment of proposed projects to evaluate and minimize the impact of development on the environment [183]. Following such work, remedial measures can be recommended to alleviate the impacts.

When used together with Remote Sensing, Geographic Information System can help to monitor the environment. It can, for example, be used to monitor land use changes [181]. It can also examine whether land development is following the land use plan of the region, by overlaying a land development map produced from the analysis of Remote Sensing images on the land use plan. In addition, Geographic Information System can be used to evaluate the impact of development on the environment to see whether adjustments of the plan are needed. A Geographic Information System can also be used in the monitoring and programming of land development [186].

6.3.5 Remote Sensing and Geographic Information System for Spatial Decision Support Systems (Sdss), and Planning Support Systems (PSS) in Town Planning

Decision support systems (DSS) were developed as a response to the shortcomings of the management information systems (MIS) of the late 1960s and early 1970s. At this time, MIS did not adequately support analytical modelling capabilities or facilitate the decision-maker's interaction with the solution process. DSS provide a framework for integrating database management systems, analytical models, and graphics, in order to improve decision-making processes. They are designed to deal with ill- or semi-structured problems, which are poorly defined and qualitative in nature. The decision support system concept was extended to the spatial context in the development of SDSS [193-195]. SDSS help decision makers to make decisions on different locational alternatives (for example, optimal location of service centres). Because of the lack of analysis functions in the past, Geographic Information System has not been considered as part of SDSS. Instead, Geographic Information System has been used to generate and store spatial data, which were then used as inputs for the analytical models. Geographic Information System then displayed the results of the analytical models. Much research has been done on the use of Geographic Information System in the visualization of the results of the analytical models. More recently, advances have been made in incorporating analytical models into Geographic Information System. For example, location–allocation and spatial interaction models have been incorporated as a standard function in ARC/INFO. It is expected that in the future, the distinction between a Geographic Information System and spatial decision support system will become increasingly blurred.

A parallel development in the planning field is the concept of a planning support system (PSS). A PSS, as first advocated by Harris [196], is a combination of computer-based methods and models that support planning functions. PSS not only serve as a decision support system to decision-makers, but also provide the tools, models and information used for planning (that is, the information technologies that planners use to carry out their unique professional responsibilities: Harris and Batty) [197]. PSS comprise a whole suite of related information technologies (e.g., Geographic Information System, spreadsheets, models, and databases) that have different applications in different stages of planning [198, 199]. A Geographic Information System are becoming important components of PSS by virtue of the geo-processing, graphic display, and database and modelling capabilities they possess. However, a PSS cannot consist of a Geographic Information System alone. It must also include the full range of the urban planning and Geographic Information System planner's traditional tools for economic and demographic analysis, forecasting, environmental modelling, transportation planning, and land-use modelling. It should also include other technologies such as expert systems, decision support aids (e.g., multi-criteria decision analysis), hypermedia systems, and group decision support systems [198].

6.3.6. Reasons and Factors Influencing the Use of Remote Sensing and Geographic Information System in Town Planning

The database management, mapping, and spatial analysis functions of Geographic Information System have been very useful in many areas of town planning [200]. The main weakness of Geographic Information System in town planning is their linkage with urban planning models. There are different strategies for linking planning models with Geographic Information System. They range from loosely coupled, to tightly coupled and fully integrated architectures. Integration based on the loosely coupled architecture involves importing and exporting data between Geographic Information System and planning models. Data are exported from a Geographic Information System and transferred to an external program for execution. The modelling results may be sent back to Geographic Information System for display and further analysis. Tightly coupled architecture integration involves writing programs within the Geographic Information System environment, avoiding explicit data transfer between software packages. Today the loosely coupled architecture is the most commonly used in the integration of planning models and Geographic Information System.

The California Urban Futures Model [189] uses the loosely coupled approach in the integration of Geographic Information System with models. It makes the use of Geographic Information System extensively for the data manipulation and displaying the results of external modelling packages. Such approach is also used in multiple criteria decision-making planning methods [201]. The tightly coupled approach has been used for residential location modelling in the Buffalo, USA region [202, 203]. Currently, there is an increasing trend toward the development of fully integrated planning models in Geographic Information System. For example, location–allocation and spatial interaction models are now in ARC/INFO. This saves the users from programming the models themselves. However, until most of the commonly used planning models are fully integrated in Geographic Information System, there is still a need for loosely coupled and tightly coupled architectures. Faust N.L. [206] discusses the issue of software coupling in a business and service-planning context by Battly [204-206] and in relation to the development of desktop GIS. Data visualization is very useful to planning. It can help planners to explore data, and calibrate and visualize the results of planning models [205, 206]. It is useful for displaying the results of complicated location–allocation models [193]. It can also enhance human–computer interaction in the decision-making process [194]. Thus, data visualization is a very valuable Geographic Information System function in a planning decision support system and the development of user-friendly data visualization functions has made Geographic Information System more useful for planning.

The integration of multimedia data types and the use of Geographic Information System in the Internet/Intranet environment (especially the World Wide Web, WWW) are some of the recent Geographic Information System developments that have had great impacts on urban planning. The incorporation of images, videos, aerial photographs, and sound within Geographic Information System can increase the planners' comprehension of the planning problem that they are analysing. The integration of Geographic Information System with virtual reality can enable planners to examine the space that they are planning more realistically. The Internet and intranets are useful for communicating design ideas [207]. The use of Geographic Information System in the Internet and environment can also facilitate the dissemination of planning information and enhance citizen participation in the planning process [208]. Unlike the past in which citizens had to go physically to the town hall to examine plans, they can now see them in their offices and homes via the Internet at any time. They can help in the development of multimedia-based collaborative planning systems [185].

The use of the Geographic Information System in town planning does not depend solely on the development of new Geographic Information System software and hardware. Indeed these considerations may be amongst the least important in influencing whether Geographic Information System is used in urban planning in a city. The status and character of the organization, data, state-of-the-art of planning, and staff are very important factors and this section considers their importance with particular regard to developing countries.

6.3.7. Application of Remote Sensing and Geographic Information System Outcomes

No matter how sophisticated and advanced it is, a decision support system is useless if decision-makers are not using it. Studies on the applications of Geographic Information System repeatedly show that staff and organizational factors are more important than technology in successful applications of Geographic Information System [192]. There are three sets of conditions that are important in effective implementation of Geographic Information System:

1. an information management strategy that identifies the needs of users and takes into account the resources at the disposal of the organization;
2. commitment to, and participation in, the implementation of any form of information technology by individuals at all levels of the organization;
3. A high degree of organizational and environmental stability [192]. Geographic Information System that is most likely to be used is those that can deal with identifiable problems. Applications that are more complex are less likely to be developed than simple ones.

6.3.8. Data Required for Town Planning

The lack of available data remains one of the major hindrances in the use of Geographic Information System [187]. As a type of information system, Geographic Information System needs graphic and textual data in order to function. There is no life in Geographic Information System without applications and there can be no application if there are no data. In short, data are vital to Geographic Information System. In the developed countries, a reasonable amount of geographical data is available, thus making the establishment of a Geographic Information System relatively easy [209] if sometimes expensive [210]. Unfortunately, data are not so readily available in the developing countries. The most readily available data are those from Remote Sensing which means that they are restricted largely to land cover information from which a very limited amount of information about land use can be extracted. Because of this, it is not surprising to find that data contained in Geographic Information System in developing countries are primarily concerned with the physical environment and land cover. Socioeconomic data, which are vital to urban and regional planning, are generally lacking and often limited only to population census data. Socioeconomic data require field surveys, which are expensive and time consuming. In a study of information, systems for planning in urban planning and Geographic Information System [211] found that there was a lack of up-to-date and reliable data for planning because some departments were slow in the acquisition of data, and coordination among various departments and institutions in the flow and exchange of information was weak. The lack of financial resources and trained personnel to collect data, and the unavailability of modern and efficient data processing equipment were also cited as inhibiting factors. The relatively low price of microcomputers can help to alleviate some of the hardware constraints in electronic data processing facilities. However, the main obstacles still lie on the failure of developing country governments to recognize the need for statistical information for planning, and their willingness and ability to mobilize resources to collect it.

In many developing countries, the base-maps, which are more essential than textual data to Geographic Information System, are often lacking or outdated. There are usually many types of base-map, each compiled by different agencies with different accuracy and map scales. This makes them difficult to integrate into Geographic Information System. Furthermore, there is a general lack of a standardized geocoding system, which makes it difficult to link the textual data and graphic data. It is, of course, difficult to start Geographic Information System without the necessary maps, geocoding systems, and textual information.

It is not only the availability of data, which is a problem in developing countries but their quality too. The currency of data is very important in planning because data will be of limited use if they are outdated. There is, unfortunately, a lack of institutional arrangements to determine, coordinate, and monitor the frequency of data updating by different departments. Large amounts of data used for planning in developing countries are collected by agencies over which the planning agency has little control. Additionally, there are no procedures for verifying the quality of the data collected.

There has been a misconception that hardware, software, and human resources are major constraints on the use of Geographic Information System in town planning in the developing countries. In fact, the availability of up-to-date data is the most important bottleneck in preventing Geographic Information System from being used effectively in the developing countries. There is an urgent need to establish an institutional framework to ensure that the required data are regularly collected and updated.

6.4. State- of- the-Art of Remote Sensing and Geographic Information System in Town Planning

The state-of-the-art in planning in the developing countries has not advanced much in comparison to Geographic Information System. The skills of planners and the planning systems may not be ready to utilize the data and functions provided by Geographic Information System. Planners could employ Geographic Information System in conjunction with new planning techniques so that they are able to diagnose potential problems and assess the desirability of alternative plans much better. In spite of this, most planners in the developing countries are not yet aware of the benefits and potential applications of Geographic Information System. Furthermore, even though much effort has been spent on data collection, comparatively little has been spent on transforming data into information for making planning decisions.

With the rapid growth of Geographic Information System, there is a shortage of human resources even in the developed countries. This shortage is more severe in the developing countries in both absolute numbers and relative terms. Very few people know about Geographic Information System in developing countries. The shortage of skilled personnel is currently very severe, especially given number of cities and regions in the developing countries that can benefit from Geographic Information System.

The problem of training is very severe in the developing countries because of the lack of expertise and shortage of funds. In developing countries, Geographic Information System is often led by teaching and research in the universities. The reverse is the case in many developing countries where, very often, it is government agencies, which buy and use GIS first through funding from international agencies. It can be several years later before Geographic Information System equipment is installed and courses are offered in the universities and other tertiary institutes. Software companies often carry out training of Geographic Information System personnel by either sending their staff to run short courses on site or sending the operators and users to be trained in their headquarters. A Geographic Information System courses and training programs if available in the developing countries are normally carried out in the departments of surveying, Remote Sensing, and geography. There is also a general lack of Geographic Information System courses in planning schools. Planners must be trained not so much on the operation of the system, but on how to make use of the data and functions of Geographic Information System in different processes of planning. At the same time to plan more effectively the planning, planners must be trained not so much on the operation of the system, but on how to make use of the data and functions of Geographic Information System in different processes of planning and plan evaluation.

A Geographic Information System is increasingly being used in planning agencies in the developed and developing countries [189]. Many planning departments that have acquired mapping systems in the past have since shifted to Geographic Information System in lieu of mapping software. With the increase in user friendliness and the number of functions of Geographic Information System software, and the marked decrease in the prices of Geographic Information System hardware, Geographic Information System is now an operational and affordable information system for planning. It is increasingly becoming an important component in the planning support system. Recent advances in the integration of Geographic Information System with planning models, visualization, and the Internet will make Geographic Information System more useful for town planning. Today, the main constraints on the use of Geographic Information System in town planning are not technical issues, but the availability of data, organizational change, and staffing.

6.5. Engineering Services in Town Planning with the Use of Remote Sensing Method and Geographic Information System Technology Applications

The progress of town planning analysis and growth with the use of Remote Sensing and Geographic Information System demands the consideration of environmental monitoring and resources investigation and outlines its achievements as well as orientation of further analysis of possibility of town planning expansion from the point of view of spatial distribution patterns and temporal change. It is very important to be aware of how Remote Sensing and Geographic Information System technologies developments can be combined integrated for the use in town planning and management process [212-214].

6.5.1. Perspectives of Town Planning Development

A city can be regarded as a hinged centre of material flows, energy flows, and information flows within a region. In other words, it is a mixed complex of social and economic development, and regional resources exploitation, continuously promoting productivity levels and the standard of living in the region. We may describe this as similar to the working of the human kidney, which constantly outputs waste products and replaces them with fresh products. Processes of urbanization reflect not only the progress and development of regions, but also mark regional cultures and civilizations of an epoch. Fating rapid change and a complicated urban system, we undoubtedly need to apply 20th century space and aerial techniques for dynamic monitoring. Furthermore, only cities can meet the needs of the high input-and-output talents and funds for Geographic Information System and Remote Sensing applications.

In 1800, the number of urban residents represented only 2 percent of the total population of the world, while the number had reached 10 percent by the beginning of 20th century. Almost two-hundred years later, the urban population has grown to over 3.2 billion, equally half of the global population, and it increases by an average of 1,000,000 people per week.

Between 1970 and 1985, the number of people below the poverty line in urban areas increased by 22 percent, so that the problem was more serious for those in urban areas. In 1993, about 330-million urban residents lived in urban areas of 16,000 km^2 in China, the numbers of urban residents have been doubled since those of the 1970s. Today there are over 660 cities with residential populations of more than 500,000.

China is both a country with a huge population and an Asian-Pacific country with a rapidly developing economy. As urbanization processes are being accelerated and more and more mega-cities are appearing, the Chinese population will increase from 1.2 billion to 1.6 billion in the 21st century. Correspondingly, the urban population will reach 29 percent in 2020 and 47 percent in 2050 [215] we will then have to face a series of environmental problems, such as a water resource crisis, increased air pollution, and waste treatment problems. We will have to make a series of decisions, such as construction planning and land management. All of these situations will stimulate the promotion of Geographic Information Systems and Remote Sensing monitoring to deal with the present and future urban problems.

Under China's extended open-door policy and the establishment of special economic regions, development zones, and bonded duty zones throughout the country, regional development in coastal areas has advanced rapidly since the 1980s.

Four highly dense urban regions have gradually formed in the Shandong Peninsula, the Liaodong (eastern Liaoning) Peninsula, the Beijing-Tianjing-Tangshan region, the Yangtse River Delta, and the Pearl River Delta. Sixteen mega-cities are distributed along 18,000 kilometres of China's coastline, including the Special Administrative Region of Hong Kong, along with 102 large-and medium-size cities and 278 small-size towns. Meanwhile, four huge, densely populated urban areas are scattered along the 6,300-kilometer Yangtse River. These include Shanghai, Nanjing, Wuhan, and Chongqing. There are, however, few cities of such size in the wider western and middle inland areas of China, with the exception of Zhengzhou, Xian, and Harbin. The trend of urban distribution mentioned above further enlarges the gap between both sides along the Tengchong-Aihui population distribution line.

In the early 1980s, aerial and satellite Remote Sensing techniques were applied to environmental monitoring, land resources investigation, and urban planning and management in China. Tianjin was the first city to conduct an aerial Remote Sensing experiment. Land-use and land-cover maps were produced, based on false-colour infrared images. Meanwhile, environmental quality evaluation was conducted on air, water source and float dust, and content of sulphur dioxide through aerosol sampling and monitoring data of air pollution. The research results were published in the Tianjin Atlas of Environment Quality [214]. This achievement had a positive effect on municipal engineering construction projects such as pollution protection of the Haihe River, crossroad planning, planting survival rate, harbour deposit monitoring, and coastal zone development with Remote Sensing data and maps. It also supplied a scientific basis for the application for a World Bank loan for municipal infrastructure construction. In 1985, NOAA AVHRR and Landsat MSS images were used to evaluate the Ecology and Environment in the Beijing-Tianjin-Tangshan region [216].

6.5.2. Remote Sensing Method and Geographic Information System Technology in Photogrammetric Engineering

In 1983, aerial Remote Sensing was successfully applied to the urban information system. According to our primary study tourism, land resources investigation, transportation, and heat [216], the trend started slowly, but developed rap islands in Beijing. The results show, for example, that is the idly in the southern coastal cities in China. Urban planning and length of Great Wall sites should be extended from 276 km to land-management information systems which have been developed to the length 673 km. The large cities in the Yangtse River Basin have sites between the second and the third ring roads where were built their urban information systems into a public benefit sup illustrated. It provided a basis for allocating seven waste treat - ported by provincial governments, with the provision of segment facilities and provided the site selection for the vices for related departments of planning and management.

Asian Olympic Game Park. For most large cities in the northern mainland of China, how- During 1985 and 1995, a series of investigations carried out for town Geographic Information System are still at the research and development stage. It has been applied with aerial infrared Remote Sensing method on ecological and/or the experimental stages. The method of Remote Sensing and Geographic Information System development are used for the environmental aspects of Taiyuan, Dalian, Guangzhou as a large- and medium-size cities in China. Their spread instance, urban land-use patterns, heat-island effects, and park- is advancing from south to north, in a fashion similar to that of land expansion rates were analysed by means of Landsat TM the modern democratic revolution and economic reforms in images. In Luoyang, Yichang, Suzhou, and other medium-size China, Regional differences remain quite high. An aerial photo and SPOT images were used for producing modern maps of selected area for urban planning.

State Land of Remote Sensing and the Global Positioning Management Bureau has established for carrying out a sustainable development strategy, with purpose of development of urbanization and enhance of land utilization. For this achievement Remote Sensing Investigations and Geographic Information System have been used in successful urban planning with proper land use in case of the large- and middle-size management processes. The statistics from 66 cities show are moving from the phase of scientific experiments to the that cities located in the eastern plains occupy huge amounts industrialization phase, and from governmental actions to the farmland while the proportions are lower in coastal cities or enterprise activities and socialization. This investigation will be Sensing Centre of China carried out research on the criteria for extended to cover more than 200 large and middle-size cities in a resource and environmental information system in 1983. The China by the year 2000 the State Standard Bureau and State Mapping and Surveying the trend in urban Geographic Information System gradually changed. Based on current circumstances an urban Bureau in 1996 has established the standard for the urban inform dynamic monitoring system for construction of a long-term facility system. Currently a number sets of monographs and teaching materials about town Remote Sensing and Geographic Information System have been available for use and its application depends on existing tasks and scope of work needed to be implemented [217, 218].

6.6. Town Planning Expansion

In the 1990s, Chinese urban Remote Sensing continued to move forward. Remote Sensing data were used to trace the course of urbanization and the history of urban evolution, and to explore the pulse of urban development. This provided the scientific basis for establishing an urban development plan. In addition, multiple Remote Sensing data were used not only as the main information source for updating the database of the urban Geographic Information System, but also as the operating system for urban dynamic monitoring.

Spatial and Temporal Analysis of Urban Growth and development of Chinese cities has a long history, and there are large regional differences. The construction of new urban areas and the reconstruction of old zones continue with unprecedented speed. Many existing environmental problems remain unsolved while more and more new problems have come into being. Therefore, the applications of Remote Sensing and Geographic Information System have become the focus of interest of both urban planning and management departments and experts of Remote Sensing and Geographic Information System who wish to gain a deeper understanding of problems such as urban-rural changes, landscape structure, and environmental quality.

As it has been called as Beijing Case, the first false-colour infrared aerial photographs were taken over Beijing in 1983. Twenty seven thematic maps were made from these photographs [219, 220] used TM images dated 26 September 1987, 17 October 1989, 25 October 1992 and 7 September 1994 to detect land-use change in Beijing. It was found that the built-up area in Beijing increased about 39 percent from 1987 to 1994. New greenbelts were established in suburban and rural areas. In 1987, parkland in Beijing and its eight satellite cities occupied & 5 percent of the total area. This increased to 16.93 percent in 1994. Obviously, this percentage is still very low. The transportation network has rapidly expanded into the suburbs. The third and fourth ring roads were legible on the TM images. In 1987 there were about 20 roads connected with these two main roads. In 1994, however, there were more than 100. Thus, traffic accidents and traffic jams increased significantly. Beijing is one of the few major world cities without a river. In the 1980s the area of public water surface was only 5.9 km^2. This went up to 20 km^2 in the 1990s, the increase being mainly the result of an increase in the number of fishponds in the area.

Research by Prof. Gu [221] revealed the trends of Beijing dynamic change. The commercial downtown areas declined and needed full reconstruction. For example, 31 supermarkets were constructed in 1995, while 60 supermarkets were reconstructed in the 1996 to 1997. Land use patterns in the inner city have been changing. In 1998, 35 towns and industrial zones were constructed in the outer suburbs, distributed in a ring whose radius extended some 20 to 70 km from the city centre. Three-hundred fifty-eight industrial enterprises inside the third ring road, which occupy 40,000 m^2, are being moved out to the suburbs in rapid succession. A large number of villas and recreational villages are being constructed in the outer suburbs.

6.6.1. Shanghai Case

The first aerial photographs were taken over Shanghai in 1947 at a scale of 1:8,000. Since 1984, aerial infrared photography surveying has been carried out ten times, at scales ranging from 1:10,000 to 1:35,000. All these photographs were widely used in the investigation of city construction; water, air, and waste pollution; urban parklands; ecological environment and land use; as well as archaeology and disaster research [222]. For example, Remote Sensing photos and images acquired in 1993-1994 were used during extant land investigation and mapping. All the results were used in the modification of master urban planning for Shanghai [223]. Through the survey and research, we obtained all the necessary data regarding construction density and dangerous house distribution. At the same time, the research analysed the geometric scales of urban roads, public parking lots, and the speed, density, and flow of vehicles. A general environmental quality assessment model of the Pudong District was established based on land use, parklands, dumping of solid waste, pollution of surface water, and distribution of feedlots. A three-dimensional simulation of urban parkland growth was conducted. The ecological and economic benefits of existing parklands were assessed. The urban heat island, as well as the type and distribution of land surface, was studied.

At the same time, the progress and regularity of Shanghai's urbanization in the past half century were studied. Using 14,000 aerial photos, ten Landsat MSS images, ten Landsat TM images, and some SPOT images, we discovered that the built-up area in Shanghai had expanded from 91 km^2 in 1947 to 364 km^2 in 1996. The rGe of increase was not uniform over the years and the region, as listed in Table 1. The rate was 3.43 km^2 per year from 1949 to 1964, 1.31 km^2 per year from 1964 to 1979, and 10.33 km^2 per year from 1979 to 1996. For example, the Pudong District expanded only 25 km^2 from 1947 to 1988 before opening, averaging 0.61 km^2 per year. After opening in 1989, the Pudong District expanded to 78 km^2 in 1993, 96 km^2 in 1996, totally 57 km^2, with an average of 8.14 km^2 per year.

6.6.2. Dongguan

Dongguan City is a case of the fastest development in the Pearl River Delta [224-226]. It was found through analysis of TM images that about 21,285.7 km2 of farmland were used for urban construction from 1988 to 1993. The TM images were also used to classify the land-use types and the changes among the different types. When evaluating land resources, the near-infrared and infrared bands were used to establish the NDVI vegetation index. The index was modified based on soil and topographic slope data, for calculating the regional agricultural suitability. Through the suitability analysis, the land resources in Dongguan were divided into seven levels, and each land parcel was assigned an agricultural loss value. Overlaying the suitability map with the administrative boundary, the agricultural loss for each town and village can be calculated.

a. Town Planning and Management Information Systems Guangzhou Urban Planning and Management Information System

The development of the Guangzhou urban planning and management information system was based on a large numbers of aerial photographs, including more than 2,000 digital orthophotomaps at 1:2,000 scale covering the city [227].

The system's purpose was to investigate hydrology and greenbelts and to identify any buildings not meeting regulations. With the support of the system, a strategic development plan for the period from 1991 to 2010 was put forward [228], including the structure of urban developmental and use patterns as well as environmental protection should be considered.

Through a comprehensive threshold analysis within a Geographic Information System, the area for urban development can be divided into different grades at certain thresholds, in order to evaluate the conditions of various development directions and the distribution of urban functions. The analysis results show that areas such as the Baiyun Mountain in Guangzhou city and the Pearl River alluvial plain with certain natural thresholds cannot be used as construction land. The existing developed area has several grades of thresholds, and great development potential will lie in the eastern part of the city.

Development of protection and planning information systems for historic cities includes information about the environmental landscape, cultural relics and historic sites, modern outstanding buildings, and old trees. There are more than 200 cultural relics and 300 old trees registered into the system.

Establishment of the database of underground pipelines covering more than 300km^2 has been established to realize office automation of pipeline planning and implementation through comprehensive query, network profile analysis, and pipeline design. A theory of six stages and a concept of three impetuses are put forward, which form the basis of the model.

Town Remote Sensing in City Chongqing became a provincial-level municipality in 1997 with an area expanding from 23,000 km^2 to 82,000 km^2, and covering 80 percent of the Three Gorges reservoir area. Chongqing is facing arduous problems such as rebuilding the old industrial bases, assisting with migration resulting from the Three Gorges Dam project, and protecting the environment. Therefore, Remote Sensing and Geographic Information System are of meat value in helping to solve the problems of city, reconstruction, farmland misuse, deforestation, and water and soil loss. 1:60,000-scale aerial false-colour infrared photography with a swath width of 15 km was taken in the area along the Yangtze River in 1984. Infrared photography at a 1:60,000 scale of city areas and 1:10,000-scale of 772 km^2 of suburban areas was taken again in from 1992 to 1995. MSS images were also used to investigate urban landforms in 1989, and the development of an urban geographic information system began in 1996. The system has been put into operation to provide basic information for regional planning and urban construction.

b. Town Development for Migration in the Three Gorges Reservoir Area

The world-renowned Three Gorges Dam is now under construction. After completion, the reservoir will cover an area of 1084 km^2. Two cities, 11 counties, 111 towns, and 1,599 factories need to be moved out. About 32,000, ha of farmland, forest, and river shoals will be submerged and 1.2 million residents will have to be re-located. In order to manage this unparalleled huge migration, satellite and airborne Remote Sensing, Global Positioning System and Geographic Information System technologies have been used to supervise the status and dynamic change information.

At the beginning of the project evaluation, land resources in the reservoir area were investigated using satellite and airborne Remote Sensing, and an atlas of the Yangtze River Three Gorges ecological environment was published. Immediately following government approval of the project in May 1993, several teams were organized to devise the plans for the emigrations that were to take place in the Three Gorges reservoir area in Fengjie, Wushan, and Badong counties, based on several time phases such as pre-closure of dam in 1997, and the second phase of emigration in 2003. The project management information system was built up based on the new generation of multi-sources Remote Sensing data, Global Positioning System and field observation. This system has been put into full use.

A new trends in urban Remote Sensing and information systems urban database was updated by using high spatial resolution satellite images. The atlas of Space View of Hong Kong, published in 1999 [229], consists of high-resolution images such as the Russian COSMOS (2 meters), French SPOT-4 (10 meters), Canadian RADARSAT (10 meters) satellite images, and Chinese scientific experience satellite images (3 meters). These images and their corresponding interpretation maps vividly depict the achievement of Hong Kong urban development programs, providing first-hand data for updating the urban database. The four new towns of Quanwan, Shatin, Tunmen, and Dapu have developed as satellite towns. Yuen Long, Fanling, and Sheung Shui are also rapidly developing. The development of these new towns provides a great amount of land, with a well-developed infrastructure for industry and business, and more space for an improved residential environment in Hong Kong. Thematic maps such as vegetation types and vegetation index, land use and land cover, nature reserves, and natural disasters were made from these images.

Aerial Three-Dimensional Remote Sensing for Dynamic Monitoring of Urban Environment and Cadastral Surveying Towards has been set up at the end of the 1960s and the Chinese central government drew up a plan for Remote Sensing technology in China. Since then, research and development of Remote Sensing sensors have been included in the National Five-Year plans. The new Remote Sensing technologies have been applied to resource exploration and environmental monitoring. For example, the dynamic three-dimensional aerial Remote-Sensing, developed by the academician Xue Yongqi and Professor Li Shukai's team, was used to acquire three-dimensional urban Remote Sensing images under the integration of technologies of Global Positioning System, laser telemetry, and imaging spectral cell matching just before the return of Macao to Chinese sovereignty in December 1999.

The Application of Resource Satellite Remote Sensing Images to Town Planning and Management in Town growth is not an isolated phenomenon. Cities and their satellite towns are influenced by and dependent upon each other. For example, the growth of a mega-city may promote or impede the development of adjacent cities. Annexation and amalgamation between cities forms group cities or city cycles. Satellite Remote Sensing again becomes a higher-level hotspot for regional town planning. Current satellite image data with medium spatial resolution will continue to play an important role.

In the early 1990s, Landsat TM and MSS images were used during town systems research in China. The images were used to calculate the populated areas in the North China plain, the Yangtze and Pearl river deltas, the Liao Dong peninsula, and the Sichuan basin. A cell-based population density map was produced. This method is especially applicable to the population census in the northwest China where the population is sparse.

In October 1999, China and Brazil cooperatively launched the BERS-1 resource satellite. The satellite has two multi-spectral scanners with spatial resolutions of 19.5 meters and 80 meters, respectively. We will use them to acquire data about the latest national town development in order to revise the layout of town systems for the twenty-first century.

6.7. Town Planning Advances with Application of Remote Sensing and Geographic Information System Technology

Urbanization is an index of transformation from traditional rural economies to modern industrial one. It is a progressive concentration of population in urban unit [230, 231]. Now, India is one among the countries of low level of urbanization. In the last fifty years, the population of India has grown two-and-a-half times, but urban India has grown nearly five times. In 2001, 306.9 million Indians (30.5%) were living in nearly 3700 towns and cities spread across the country, and it is expected to increase to over 400 million and 533 million by 2011 and 2021 respectively. Now, India is among the countries of low level of urbanization. As a result, most urban settlements are characterized by shortfalls in housing and water supply, urban encroachments in fringe area, inadequate sewerage, traffic congestion, pollution, poverty, and social unrest making urban governance a difficult task. The high rate of urban population growth is a cause of concern among India's urban and town planners for efficient urban planning. For this, the government of India has taken an important initiative to strengthen municipal governance, which is the enactment of the Constitution (74th Amendment) Act (CAA), 1992. Through this initiative, an attempt is being made to improve the performance ability of municipalities/urban local bodies, so that they would be able to discharge their duties efficiently in the planning and development of urban areas.

However, most studies undertaken to assess the functioning of municipalities in India point out that the municipalities are confronted with a number of problems, such as, non-availability of data, ineffective participation in the decision-making process despite of the adoption of the policy of reservation, delays in the transfer of funds to the municipalities, the constitution of State Finance Commissions, poor recovery from various tax and non-tax sources, devolution of power etc. Therefore, there is an urgent need to adopt modern technology of Remote Sensing, which includes both aerial and satellite based systems, allowing us to collect many physical data rather easily, with speed, on repetitive basis. As well as together with Geographic Information System, it helps us to analyse the data spatially, offering possibilities of generating various options (modelling), thereby optimizing the whole planning process. These information systems also offer interpretation of physical (spatial) data with other socio-economic data, and thereby provide an important linkage in the total planning process and making it more effective and meaningful.

Planning is a widely accepted way to handle complex problems of resources allocation and decision-making. It involves the use of collective intelligence and foresight to chart direction, order harmony, and make progress in public activity relating to human environment and general welfare. Necessary support of the organization has become essential in order to provide more effective and meaningful direction for better planning and development. Hence the need for a suitable information system is increasingly being felt in all planning and developmental activities, whether these are for urban or rural areas. Urban areas of today are more exactly described as sprawling regions that become interconnected in a dendritic fashion (Carlson and Arthur, 2000 [232-234]. The positive aspects of urbanization have often been overshadowed by deterioration in the physical environment and quality of life caused by the widening gaps between supply and demand for essential services and infrastructure.

Urbanization is inevitable, when pressure on land is high, agriculture incomes are low and population increases are excessive, as in the case of most of the developing countries of the world. Urbanization has become not only of the principal manifestation but also an engine of change in urban processes. An urban transition period for human society has become in the 21th century. In a way, urbanization is desirable for human development. However, uncontrolled urbanization has been responsible for many of the problems, our cities' experiences today, resulting in substandard living environment, acute problems of drinking water, noise and air pollution, disposal of waste, traffic congestion etc. To improve these environmental degradations in and around the cities, the technological development in relevant fields have to solve these problems caused by rapid urbanization, only then the fruits of development will reach most of the deprived ones.

As it is said above, the modern technology of Remote Sensing which includes both aerial and satellite based systems, allowing us to collect a lot of physical data rather easily, with speed, on repetitive basis. As well as together with Geographic Information System it helps us to analyse the data spatially, offering possibilities of generating various options (modelling), thereby optimizing the whole planning process. These information systems also offer interpretation of physical (spatial) data with other socio-economic data, and thereby provide an important linkage in the total planning process and make it more effective and meaningful.

Recent technological advances made in domain of spatial technology cause considerable impact in planning activities. This domain of planning is of prime importance for a country like India with varied geographic patterns, cultural activities etc. The purpose of using Geographic Information System is that, maps provide an added dimension to data analysis, which brings us one-step closer to visualizing the complex patterns and relationships that characterize real-world planning and policy problems. Visualization of spatial patterns also supports change analysis, which is important in monitoring of social indicators. This in turn should result on improving need assessment.

6.7.1. Theoretic Aspects of Town Planning

The urban is a compound system of human and nature. It is also a high-dense geographical synthesis of population, resources, environment, and social economic and so on. As one sign of civilization and social progress, the city's effects on national politics, economics and culture become prominent increasing. In other words, the urbanization's level is a significant parameter to measure a country's extent of civilization, social progress and economic. Therefore, it is very important to make reasonable and fit urban planning and management [235, 236].

According to the urban development aims, urban planning constitutes the urban character, scale and development direction make use of the urban land reasonably. Town planning relates to politics, economic, society, technology, art and comprehensive domains of human life. It is not only integrated, but also concerned with the policy and practice [237-239].

The primary phase of urban planning is present situation investigation. In the past, it always consumed a lot of labour force, material and money. The result was not timely and exact. Nowadays, the Remote Sensing technology can be used to investigate urban terrain, physiognomy, lakes, plants, sights, traffic, land utilization, and building distribution quickly. As a main method to obtain and update urban geometric information and some attribute information, the Remote Sensing technology is quick, exact, and economical [240, 241].

6.7.2. Factors of Town Planning

India's urbanization is often termed as over-urbanization, pseudo urbanization, because not due to urban pull but due to rural push. The push factors like population growth and unemployment etc. (demographic factors) and pull factors like opportunities (economic factors) in the urban area. The globalization, liberalization, privatization are addressing negative process for urbanization in India. The big cities attained inordinately large population size leading into virtual collapse in the urban services and followed by basic problems in the field of shortfalls in housing and water supply, inadequate sewerage, traffic congestion, pollution, poverty and social unrest making urban governance a difficult task. For this, the government of India has taken an important initiative to strengthen municipal governance, like the enactment of the Constitution (74th Amendment) Act (CAA), 1992. Through this initiative, an attempt is being made to improve the performance ability of municipalities/urban local bodies, so that they are able to discharge their duties efficiently in the planning and development of urban areas. Urban Local Bodies [ULBs] that are statutorily responsible for provision and maintenance of basic infrastructure and services in cities and towns are under fiscal stress. According to Census of India 2001, there were 5621 ULBs in the country classified into three major categories of municipal corporations (500), municipalities (50-500) and town committees (5-50).

The 74th Constitutional Amendment Act (CAA74) mandates compulsory reconstitution of municipal bodies within a stipulated period, thus ensuring continuity of local representatives. The twelfth schedule (Article 243W) of the CAA74 has listed 18 functions and responsibilities to local bodies. These are:

1. Urban planning, including town planning;
2. Regulation of land use and construction of buildings;
3. Planning for economic and social development;
4. Roads and bridges;
5. Water supply for domestic, industrial, and commercial purposes;
6. Public health, sanitation, conservancy, and solid waste management;
7. Fire services;
8. Urban forestry, protection of the environment, and promotion of ecological aspects;
9. Safeguarding the interests of weaker sections of society, including the handicapped and mentally retarded;
10. Slum improvement and up-gradation;
11. Urban poverty alleviation;
12. Provision of urban amenities and facilities such as parks, gardens, and playgrounds;
13. Promotion of cultural, educational and aesthetic aspects;
14. Burials and burial grounds; cremation grounds and electric crematorium;
15. Cattle pounds, prevention of animals from cruelty;
16. Vital statistics, including registration of births and deaths;
17. Public amenities including street lighting, parking lots, bus-stop, and public conveniences;
18. Regulation of slaughterhouses and tanneries.

Importantly the 12th schedule of CAA74 expressly recognizes a role for the ULBs within the constitutional framework and provides devolution of financial powers from the state government for strengthening of municipal finances. The CAA74 also provides for constitution of Ward Committees in municipalities with a population of more than three lakhs, Metropolitan Planning Committees and District Planning Committees for consolidation and preparation of plans of spatial, economic, and social development. From a "top down" approach, the emphasis has thus shifted to the "bottom up" approach. In view of the challenges facing by ULBs, the planners have to prepare themselves for a new role and much wider responsibilities. As a bridge between the civil society and the politico-economic structure, the planners have to perform the role of the catalysts of change. With the ongoing globalization, economic liberalization and devolution of power to local bodies have been diluted armchair of professionals' activities.

6.7.3. Town Planning Requirements

A review of the attempts made for introducing Remote Sensing and GIS based urban planning practices brings forth a set of problems coming across with our planning organizations. They are highlighted below:

Technical

- Lack of appropriate base maps necessary for micro-level and utility planning.
- Difficulty in correlating Remote Sensing data with corresponding cadastre information.

- Limitation on availability and digitization of certain data products.

Financial

- Inadequate funds to acquire and upgrade periodically the hardware and software.
- Absence of provision for repair and maintenance service due to which upkeep of hardware suffers.
- Inability to procure digital data products and carry out surveys for collection of attributed data.

Institutional

- Absence of a dedicated team that would continue for a reasonable period to establish GIS database.
- Tendency to hold on the information due to which GIS database creation cost is not shared.
- Lack of support to young GIS professionals by the peers who feel threatened.
- Rigidity in work culture not encouraging experimentation that is so vital.

6.7.4. Remote Sensing and Geographic Information System Implementation

Most of these problems have their origins in the fact that urban planning falls under the public sector namely State Government and Urban Local Bodies whose limited financial resources and capacity to innovate do not help the cause of the Geographic Information System. However, late various schemes of the Government of India, promotion of public-private sector joint ventures and interest shown by many international agencies for collaboration in the field of geo-informatics has brought some changes to the situation. Since the problems are identified, it would not be impossible to overcome them, especially since the power of Geographic Information System and Remote Sensing in the field of urban planning is well recognized.

a. Stages of Town Planning

Urban areas face many critical environmental problems, which are manifested at the time of crisis. To avoid such occurrences, the prime requirement is quantification or "resource potentiality," its availability and consumption in the urban areas, which requires a comprehensive Urban Information System (UIS) to be developed to cater the developmental needs of the growing urban areas.

- Thematic map preparation from satellite data using visual interpretation techniques;
- Generation of spatial framework in Geographic Information System environment for perspective and development plans;
- Integration of thematic maps using Geographic Information System techniques for urban sprawl analysis and urban land use change analysis;
- Area required for urbanization will be determined on the basis of population projection of the city and its growth centres;
- Calculation of land requirements for urban development based on the carrying capacity of the region;
- Projection of urban land use suitability analysis;

- Urban environmental sensitivity analysis based upon both physical and air quality parameters;
- Determination of composite functionality index to setup various amenities such as educational, medical, recreational, etc.

b. Remote Sensing and Geographic Information System Applications in Town Planning

In India, the complexity of urban development is so dramatic that it demands immediate attention and perspective physical planning of the cities and towns (**Sokhi and Rashid**, 1999) [242]. The dynamic nature of urban environment necessitates both macro and micro level analysis. Therefore, it is necessary and fundamental for policy makers to integrate like Remote Sensing into urban planning and management. Traditional approaches and technique designed for towns and cities may prove to be inadequate tools when dealing with metropolis. New approaches are required, and new methods must be incorporated into current practice. Until recently, maps and land survey records from the 1960's and 70's were used for urban studies, but now the trend has shifted to using digital, multispectral images acquired by EOS and other sensors. The trend towards using Remotely Sensed data in urban studies began with first-generation satellite sensors such as Landsat MSS and was given impetus by a number of second-generation satellites: Landsat TM, ETM+ and SPOT HRV. The recent advent of a third generation of very high spatial resolution (5 meter/pixel) satellite sensors is stimulating. The high resolution PAN and LISS III merged data can be used together effectively for urban applications. Data from IRS P-6 satellites with sensors on board especially LISS IV Mono and Multispectral (MX) with 5.8 meter/pixel spatial resolution is very useful for urban studies.

Advancement in the technology of Remote Sensing has brought miracle in the availability of the higher and higher resolution satellite imageries. They are: IRS-P6 Resource sat imagery with 5.8 meter resolution in multispectral mode, IRS-1D Pan image with 5.8 meter resolution, Cartosat-I imagery of 2.5 meter resolution with stereo capabilities, Cartosat-II with 1 m, IKONOS imageries of Space Imaging with 4 meter in multispectral mode and 1 meter in panchromatic mode, Quickbird imagery of Digital Globe with 61 cm resolution in panchromatic mode and so on. These high resolutions of the sensors provide a new methodology in the application of newly raised technical restrictions.

Apart from Cartographic applications, P-6 data will be useful in cadastral mapping and updating terrain visualization, generation of a national topographic database, utilities planning and other Geographic Information System applications needed for urban areas. The satellite will provide cadastral level information up to a 1:5,000 scale, and will be useful for making 2-5 meter contour map (NRSA 2005). The output of a Remote Sensing system is usually an image representing the scene being observed. Many further steps of digital image processing and modelling are required in order to extract useful information from the image. Suitable techniques are to be adopted for a given theme, depending on the requirements of the specific problem. Since Remote Sensing may not provide all the information needed for a full-fledged assessment, many other spatial attributes from various sources are needed to be integrated with Remote Sensing data. This integration of spatial data and their combined analysis is performed through GIS technique. It is a computer-assisted system for capture, storage, retrieval, analysis and display of spatial data and non-spatial attribute data. The data can be derived from alternative sources such as survey data, geographical/topographical/aerial maps, or archived data. Data can be in the form of locational data (such as latitudes/longitudes) or tabular (attribute) data. A Geographic Information System techniques are playing an increasing role in facilitating integration of multi-layer spatial information with statistical attribute data to arrive at alternate developmental scenarios.

Application of Remote Sensing technology can lead to innovation in the planning process in various ways:

1. Digitization of planning base maps and various layout plan has facilitated updating of base maps wherever changes have taken place in terms of land development etc.;
2. Digital maps provides flexibility, as digital maps are scale free. Superimposition of any two digital maps, which are on two different scales, is feasible. This capability of digital maps facilitates insertion of fresh survey or modified maps into existing base maps. Similarly superimposition of revenue maps on base maps with reasonable accuracy is great advantage compared with manually done jobs;
3. Since information and maps are available in digital format, correlating various layers of information about a feature from satellite imagery, planning maps and revenue maps is feasible with the help of image processing software like ERDAS Imagine, ENVI and PCI Geomatica, ILWIS. Such super imposed maps in Geographic Information System software like Map info, Geomedia, Arc View, Auto CAD Map and Arc GIS provide valuable information for planning, implementing and management in urban areas.

Remote Sensing techniques are extremely useful for change detection analysis and selection of sites for specific facilities, such as hospital, restaurants, solid waste disposal and industry. An attempt has been made here to demonstrate the potentials of Remote Sensing techniques in base mapping, land-use and land-cover mapping, urban change detection and mapping, urban infrastructure and utilities mapping, urban population estimation, management.

6.7.5. The Use of Aerospace Data in Town Planning

Aerial photographs have long been employed as a tool in urban analysis [243, 244]. In India, city planning has been largely confined to aerial photography. It is being used for generation of base maps and other thematic maps for urban areas as it is proved to be cost and time effective and reliable. Wealth of information pertaining to land features, land use, built up areas, city structure, physical aspects of environment etc. is available from the aerial photography. Various types of cameras and sensors black and white, colour, colour infrared are used for aerial photography. Because of security concerns related to aerial photography, the use of photogrammetric techniques was confined to smaller cities.

Aerial photographs provide information that can significantly improve the effectiveness of city and town planning and management in India. They are also relatively low in cost, accurate, reliable and can be obtained on desired scale, but they are not useful in large metropolitan areas. As discussed above, India very much depends on photogrammetry for providing information for urban planning purposes. But since the March 17, 1988 launch of its first satellite (IRS1A) equipped with the LISS-I sensor acquiring 72.5 meter/pixel data, the application of Remotely Sensed data (from various sensors) in urban and regional planning processes has gained momentum. LISS-I gathered data in four spectral bands (0.45 μm - 0.86 μm) and was mainly used for broad land-use, land-cover, and urban sprawl mapping. The IRS-1C and 1D satellites launched in 2003, carrying LISS-III and LISS-IV sensor with spatial resolutions of 23.5 meter/pixel and 5.8 meter/pixel using Landsat MSS optical bands (0.52 μm - 0.86 μm), have contributed to the effectiveness of urban planning and management.

Early experiments with the first generation satellites found the data very useful for mapping large urban parcels and urban extensions. The development of Landsat TM data with 30 meter/pixel spatial resolution has helped in mapping Level-II urban land use classes. Cities and towns in India exhibit complex land use-patterns, with the size of urban parcels varying frequently within very short distance. The extraction of urban information from Remotely Sensed data therefore requires higher spatial resolution.

6.7.6. Town Mapping

Base map, a pre-requisite for urban planners, refers to the large-scale maps, which depict broad physical and cultural features. The base maps are produced at a scale ranging from 1:10,000 to 1:4,000 and 1:1,000/1:500 for specific urban applications (shown in Table 14). Base maps at a scale of 1:4,000 prepared by ground survey in 1969 and 1971 are available for some areas. Now base maps are being produced at scales ranging from 1:4,000 to 1:10,000 depending upon the specific urban applications for which they are prepared.

Table 14. Town planning stages and base map requirements

N	Planning stage	Base map scale
1	Master Plan / Land use Plan	1:10,000 & larger
2	Zoning Plan	1:4,000
3	Inner City/Urban Cadastre	1:1,000 to 1:2,000
4	Urban slums/Unauthorized Developments/ Encroachments	1:5,000 to 1:1,000

Base maps can be also made from orthophotographs for inaccessible areas that are difficult to survey, high altitude towns like Leh, Puri, Himachal Paradesh etc. In such situations, Remote Sensing has made information collection possible for base maps where field surveying has fallen shot due to prohibitive faction such as cost, timing and terrain. These base maps can provide the backbone for development of information that was previously unavailable to the community, urban regional, and natural resource planners and management. IRS P-6 (multispectral) data with 2.5 m/pixel spatial resolution can meet the ever-growing demand for current, accurate base maps at a scale of 1:5,000 for urban planning purposes and for development new residential sites.

6.7.7. Land -Use and Land -Cover Mapping

Land is one of the prime natural resources. Urban population growth and urban-sprawl induced land use changes coupled with industrial development are resulting in unplanned use as well as misuse of land leading to conversion of usable land into wastelands. The changes of land-use/land-cover pattern over a period control the pressure on land [245, 246].

The complexity of urban development is so dynamic that it calls for an immediate perspective planning of cities and towns [247]. For a sustainable use of the land, it is essential that proper planning and monitoring have been done. Timely and accurate information on the existing land-use/land-cover pattern and its spatial distribution and changes is a prerequisite for planning, utilization and formulation of policies and programs for making any micro and macro-level developmental plan. Accurate, reliable and comprehensive spatial-temporal information on land use practices in a city is prerequisite for sustainable land management. Remote Sensing offers cost-effective solutions to city planners' data needs for both macro and micro level analysis of the land use planning leading to urban environment management. The better management and rational use of land is demanding for accurate and timely changes in the dimension, nature, and spatial balance between exploitation and regeneration. A Geographic Information System is the best utilized for integration of various data sets to obtain homogeneous composite land development units which helps in identifying the problem areas and suggest conservation measures.

The Remote Sensing technology along with Geographic Information System is an ideal tool to identify, locate and map various types of lands associated with different landform units [248-250].

The timely information about the changing pattern of land use plays a significant role in land use planning and sustainable land development. The mapping and monitoring of the land use/land cover requires a land use classification of a system. One of the most widely used data format for information extraction about the land-use and land-cover is the infrared False Colour Composite (FCC) image. The extraction of information from such images about ground reality is done by image interpretation for which generally three methods namely photo interpretation, spectral analysis and data integration are used. Prasad and Sinha (2002) describe the image characteristics and visual interpretation techniques of various land-cover and land-use categories, which is summarized in the Table 15 [251].

6.8. Dynamic Change Definition in Town Planning

Land-use change detection and mapping require high-resolution imageries to obtain detailed information and multispectral optical data to make fine distinction among various land-use classes. While many urban features can be detected on radar and other imagery visible and near infrared, (VNIR) data of a high-resolution permit, fine distinction among subtle Land - Cover/Land - Use classes.

Table 15. Land - Cover/Land - Use and their image characteristics Land - Cover/Land - Use image characteristics

Land-cover/land- use	Image characteristics
Settlements	Light grey clustering with particular patterns for the urban area. There may be brownish maroon patches for in between vegetation. For the rural settlement there occur no particular patterns of such image characteristics.
Agriculture	Identify rabbi if the month of data acquisition is January or February or March and colour is brown red. (a) For the Khari crops same characteristics in image occur if the image data are acquired in the month of September, October or November. (b) Fallow land is identified by light grey colour within cropped area (red colour). (c) Plantation occurs as brownish maroon patches.
Forest (a) Dense forests (b) Degraded forest (c) Forest blank (d) Forest plantation	Dense forests are identified by dark red colour patterns. In the case of degraded forest the dark red colour patterns contain small brown or white patches. The blanks in the forest show creamy patches in the dark red/background. Forest plantations are identified by dark red colour sign of particular pattern.
Waste Land (a) Muddy water logging (b) Clear water logging (c) Temporary water logging (d) Permanent water logging (e) Marshy area water logging (f) Gullied land (g) Land with scrub (h) Land without scrub (i) Sandy area	Muddy water logging occurs as blackish or deep blue spots while clear water logging area is identified by dark/bright blue patches. Comparing the images of rainy season and out of rainy season identifies temporary and permanent water logging. Marshy area is recognized as a sign of vegetation (red/pink spots) in the water logged (blackish blue/bright blue) area. Gullied land occurs as white/grey spot. The image of land with scrub contains white patches in the land area. Sandy area is classified as bright white coloration along the course of river.
Water bodies (a) River/stream (b) Canal (c) Lake/ reservoirs (d) Embankments	River/stream is identified as long non-linear path coloured with dark blue/bright blue line in white background. Canals are identified as line segments sign of water bodies. Lake/reservoirs are identified as patterns along the river. Embankment occurs as light grey structure along the river.
Others	Grasslands are identified as uneven appearance characterized by red (light to medium grey tones) Snow is identified as white patches on the hills.

Urban areas are highly dynamic. Remote Sensing can enable urban planners and decision makers to assess land-use conversions from agricultural to non-agricultural (i.e. residential, commercial, and industrial), loss of greenery and water bodies, development along main transport routes and drainages lines, and changing quality of the urban environmental. Detection of large-scale conversion of agriculture land into non-agricultural land has been useful for determining the extent of built-up areas on the day the government decides to acquire land [252].

It is also useful in dealing on urban change detection using Survey of India topo sheets at 1:50,000, Landsat MSS, IRS LISS-I and II, and SPOT HRV data. With the use of IRS LISS-I False Colour Composite (FCC), the transformation of agricultural land into residential and industrial land was carried out by the Kukatpally municipal area in Hydrabad city in 1990. Land-use/land-cover change detection mapping of Delhi has been done in ERDAS Imagine Software and Arc GIS using Landsat TM and IRS LISSIII data from 1992 and 2004 respectively. Urban change-detection mapping can be done with digital image processing (DIP) software using satellite data from different periods [253].

6.8.1. Town Enhancement

In India, unprecedented population growth coupled with unplanned developmental activities has led to urbanization, which lacks infrastructure facilities. This also has posed serious implications on the resource base of the region. The urbanization takes place either in radial direction around a well-established city or linearly along the highways. This dispersed development along highways, or surrounding the city and in rural countryside is often referred as sprawl [254]. The direct implication of such urban sprawl is the change in land use and land cover of the region. The ability to service and develop land heavily influences the economic and environmental quality of life in towns. Identification of the patterns of sprawl and analyses of spatial and temporal changes would help immensely in the planning of proper infrastructure facilities.

Patterns of sprawl and analyses of spatial and temporal changes could be done cost effectively and efficiently with the help of spatial and temporal technologies such as Geographic Information System) and Remote Sensing along with collateral data (such as Survey of India maps, etc.). IRS-1C/1D/P4 provides data with good spectral resolution (LISS data) and the spatial resolution of 5.6 m in panchromatic mode. The Remote Sensing satellites with high-resolution sensors and wide coverage capabilities provide data with better resolution, coverage and revisit to meet the growing applications needs. The image processing techniques are also quite effective in identifying the urban growth pattern from the spatial and temporal data captured by the Remote Sensing techniques. These aid in delineating the specific growth patterns of sprawl which could be linear or radial or both. The spatial patterns of urban sprawl over different periods, can be systematically mapped, monitored and accurately assessed from satellite data (Remotely Sensed data) along with conventional ground data [255]. Mapping urban sprawl provides a "picture" of where this type of growth is occurring, helps to identify the environmental and natural resources threatened by such sprawls, and to suggest the likely future directions and patterns of sprawling growth. However, the physical expressions and patterns of sprawl on landscapes can be detected, mapped, and analysed using Remote Sensing and Geographic Information System [256] with image processing and classification. The patterns of sprawl are being described using a variety of metrics and through visual interpretation techniques. Ultimately, the power to manage sprawl is residing with local municipal governments that vary considerably in terms of will and ability to address sprawl issues.

[257] bring out the techniques for mapping suburban sprawl. They evaluate the traditional unsupervised classification and proposed GIS buffering approach for mapping the suburban sprawl, and Yeh and [258] use Shannon's entropy, which reflects the concentration of dispersion of spatial variable in a specified area, to measure and differentiate types of sprawl. This measure is based on the notion that landscape entropy or disorganization increases with sprawl. The urban land uses are viewed as interrupted and fragmented previously homogenous rural landscapes, thereby increasing landscape disorganization [259] have also employed a similar approach of characterizing urban sprawl for Hyderabad city, India.

6.8.2. Town Infrastructure and Utility Mapping

The civic amenities like potable water for domestic consumption, educational institutions, recreational sites, power plants, means of transportation and waste disposal sites form some of the major urban infrastructure and utility services. The urban planners need large volume of data both at pre-planning and plan implementation stages to ascertain the status of the available facilities and to determine the actual/projected demands for the same. The Remote Sensing techniques provide accurate, orderly, and reliable information repetitively for planning and management of urban utility services. Whereas the aerial photography at a scale of 1:10,000 and larger provides information about the spatial distribution of most of the urban infrastructural facilities, the SPOT and IRS 1C & 1D data in panchromatic mode offers capabilities for mapping and analysing urban transport network, effluent discharge zones and urban greenery.

The large-scale aerial photography was used in making the quantitative assessment of habitant and its refuse and in identifying disposal sites of Kanpur [230, 231]. Similarly, an assessment of the population served by the urban facilities and services in Bhubaneshwar was studied using 1:8,000 scale photography and USEMAP-4, Geographic Information System software [260]. The SPOT MS and PLA data have been used in evaluating urban land use-transportation system relationship along Ring railway transport system in Delhi. Konkan railway and pipeline routing have also been studied using the satellite data.

6.8.3. Town Hydrosystem

The urban agglomerations in India are facing at least four hydrological problems, i.e., the mobilization of sufficient volume of water for domestic and industrial consumption, urban water pollution and quality, flood control and urban storm water run-off disposal. Indian cities face problems of insufficient water for domestic & industrial purposes, poor water quality & inadequate urban storm water runoff disposal. However, Runoff cannot be directly measured by Remote Sensing techniques. The role of Remote Sensing in runoff calculation is generally to provide a source of input data or as an aid for estimating equation coefficient and model parameters. The Remote Sensing techniques are also being applied in obtaining information pertaining to surface water quality parameters, soil, drainage, land-use, ground water, and slope of catchment or watersheds relevant to carry runoff and water estimation studies. For example, a Remote Sensing based approach was evolved to deal with the metro water supply problems of Madras. Similarly, Chakraborty discussed an approach for urban storm water, runoff modelling, water supply assessment and water quality surveillance of Delhi Urban Complex, Najafgarh, Patna, and Hydrabad (NRSA-TR, 1989). The operational utility of Remote Sensing techniques in water resources assessment of Hyderabad city has been dealt with using Landsat TM and IRS LISS-I and II data. The attempt has also been made to identify and delineate different hydro geomorphological units in and around the immediate environs of Jhansi city and correlate them with the well yields using Landsat TM FCC. Remote Sensing can be applied to drainage studies using proxies or surrogates. Satellite data have been successfully used to map surface drainage pattern.

A Geographic Information System technology has the ability to capture, store, manipulate, analyse and visualize the geo-referenced data. On thc other hand, hydrology is inherently spatial and distributed hydrological models that have been large data requirements. The integration of Geographic Information System and hydrology involves three major components:

- Spatial data construction;
- Integration of spatial model layers;
- Geographic Information System and model interface.

A Geographic Information System technique have been applied for groundwater vulnerability mapping with the DRASTIC model in Aligarh city, using the weighted sum overlay method. The DRASTIC model considers seven groundwater parameters, depth of water, net recharge, aquifer media, soil media, topography, impact of vadose zone & hydraulic conductivity. Studies on hydrology have been conducted by using a Geographic Information System - based on solution for watershed analysis of Maryland State. They constructed a model named as GISHYDRO. [261] has developed automated geo-spatial watershed assessment tool (AGWA) in Geographic Information System for analysing the water resources. The key components of AGWA are the hydrological models used to evaluate the effects of land-cover and land-use on water response.

6.8.4. Town Traffic Management System Development

The transportation network is an important infrastructure element of the whole urban area. It allows connectivity and movement of people, traffic and goods both within and between urban centres. Radar RS data can be used effectively for urban transportation network management. All roads with a width of 3m or more can be seen on high-resolution (IKONOS) satellite data; such data facilitate the identification of roads that need to be widened to ease congestion. Using satellite images, road information can be updated and the approximate width of a road can be determined. Road width can be assessed using data from SPIN-2 with 2 m/pixel resolution. ADEOS multispectral data with 16 m/pixel resolution and LANDSAT TM with 30 m/pixel resolution. A 5 m wide road can be measured with a maximum error of 1m using SPIN-2 data. A 35m wide road varies from 34m-36m in SPIN-2 data giving a maximum error of 1m. Width of the same road or road section varies from 32m - 40m & 30m - 40m in ADEOS pan data and SPOT pan data respectively, with a maximum error of 5m in each case. The effects of urban traffic on the environment in Jaipur, in terms of population affected by air and noise pollution, was studied using predictive and dispersion models in a GIS environment using 1998 data from IRS-1C, LISS III, FCC and PAN. The study showed that a significant percentage of the population was affected by air (94.3%) & noise (34.8%) pollution. About 52% of the total population residing in a 0- 425m buffer zone was affected by all air pollutants and 41.6% of the total population living in 425-1500m buffer zone was affected by suspended particulate matter. Such data are vital for formulating strategies to mitigate traffic related air and noise pollution hazards, such as mass transit, telecommunicating, and enacting stricter automobile emission standards.

6.8.5. Town Waste Management System

Solid waste is a potential nightmare for India's large and growing population, due to inadequate and legislative instruments and to the deplorable organizational and financial capabilities of local urban governments. The informal sector should be organized and the private sector should participate more widely in collection and recycling of solid waste.

In this context, the most acceptable strategy for solid waste management is first to categorize waste streams as biodegradable, non-degradable and recyclables. Then the problem is where to dispose of it and it is not easy to locate the disposal site. A geospatial database generated from Remotely Sensed satellite data could be used to help the solution of this problem. Efforts should be made to reclaim abandoned landfill sites. Attention should be focused on identification of suitable new sanitary landfill sites to isolate waste from human society and the ecosystem and monitoring of the existing landfill sites for environmental impact assessment. RS data can aid in identification and location of such landfill sites and in monitoring the changes in land-use within and near hazardous waste and sanitary landfill. Moreover, these data integrate with Geographic Information System, has been useful to identify potential waste disposal sites using IRS, LISS IV, PAN imagery with 5.8 m/pixel resolution and ASTER visible to near IR data with 15 m/pixel resolution. Path optimization can be carried out by using a network analysis model in GIS for a solid waste dumping.

6.9. TOWN PLANNING MODIFICATION

For a more dynamic urban planning exercise, the following modifications in the planning approach are recommended:

(i) Flexibility: Plans must have flexibility to provide for ever-growing and ever-expanding city boundaries and provide quality of life to all inhabitants. The plan should be flexible to respond not only to the present needs but also to the changing conditions in foreseeable future;

(ii) Role of Actors: People's participation in preparation of policies, perspective plan, development plan and annual plans should be ensured through elected representatives in the municipal council/corporation and ward committees;

(iii) Information system: A well maintained information system can make possible the fine-tuning of the plan proposals at the various stages of implementation of the plan according to the changing urban scenario;

(iv) Urbanized Areas: The development potential may be assessed for the areas located in the periphery of the developed areas. A profile of the development potential and the possibility of optimizing the existing infrastructure should determine the prioritization of development of these areas;

(v) Growth Centres: Given the paucity of resources, it would be more feasible and desirable to promote strategic development initiatives in the selected secondary cities, growth centre and their hinterlands. In the growth centres, the location of infrastructural and environmental services could form the core of the Development Plan;

(vi) Policy Guidelines: Policy guidelines notified under law, can help in identifying priority areas, subsequent modifications in the plans and administration in general;

(vii) Mixed Land Use: With a view to provide for development, the zoning regulations need to be simplified. The land use package should not be allowed to be changed by any authority, except as a part of the review of the Development Plan at the city/town level;

(viii) Financial Planning: Land development and infrastructure investment need to be coordinated through integration of physical, financial and investment planning. There is a need to link spatial development plan with resource mobilization plan focusing on credit enhancement mechanisms;

(ix) Land Policy and Management: As opposed to the process of compulsory land acquisition, and the related issue of low compensation rates, the ULBs should adopt collaborative approaches within the existing legal framework;

(x) Legal Framework: Plan implementation would call for a legal framework to make it enforceable and mandatory. The legal framework has to be supported by an effective and efficient machinery which would see that no distortion of master plan proposals takes place at the ground level;

(xi) Standards: Plot sizes, layout and social overheads need to be designed to reduce costs aligned to the affordability of different income groups and also the sale price for lower income groups can be reduced by differential pricing;

(xii) Building Byelaws: Building byelaws and zoning regulations for the city/town should match the local needs. However, the existing byelaws need to be simplified and transparent, and there should not be an element of discretion. Adequate provision for parking facilities should be made;

(xiii) Database at Metropolitan, district and state levels: The planning exercise need continuous data collection, analysis interpretation and updating of data. A computer-generated database and information system in Geographic Information System environment should be developed at various levels, which would provide support to planners in the development of planning.

Planning and managing cities in the new era of globalization and economic liberalization would be a demanding task calling for new skills and approach. Indian cities will have to compete with others to attract investments and, therefore, issues like quality of infrastructure, energy efficient services provision, and environmental conditions in a city besides economic stability would play a significant part in such competition. Urban planning profession in general will have to address these issues and respond rapidly. It is worthwhile noting that spatial dynamics of cities is complex to fathom and urban theory is still static. In other words, the urban planning authorities and agencies in every parts of the country should adopt new technologies like Remote Sensing and Geographic Information System. These have capability to provide necessary physical input and intelligence for preparation of base maps, for planning proposals and act as monitoring tool during implementation phase(s). Satellite Remote Sensing with repetitive and synoptic viewing capabilities, as well as multispectral capabilities, is a powerful tool for mapping and monitoring the ecological changes in the urban core and in the peripheral land-use planning, will help to reduce unplanned urban sprawl and the associated loss of natural surroundings and biodiversity.

On the other hand, moving further, interfacing of urban planning models with Geographic Information System should now receive due attention. Incorporation of land-use transportation models, water distribution network analysis, and simulation of urban activities to evaluate different urban development alternatives in the Geographic Information System framework needs to be explored for added advantage.

6.10. Geo-Information System in Town Planning

Urbanization is an index of transformation from traditional rural economies to modern industrial one. It is a progressive concentration of population in urban unit. During the last fifty years the population of India has grown two and a half times, but urban India has increased nearly five times. In 2001, 306.9 million Indians (30.5%) were living in nearly 3,700 towns and cities spread across the country, and it is expected to increase to over 400 million and 533 million by 2011 and 2021 respectively. These information systems also offer interpretation of physical (spatial) data with other socio-economic data, and thereby provide an important linkage in the overall planning process and making it more effective and meaningful [263].

The majority of the world's population now resides in urban environments and information on the internal composition and dynamics of these environments is essential to enable maintenance of certain standards of living.

The availability of urban land cover data is critical to policy makers, particularly for town planners, because of their ability to monitor impact of planning policies, the direction of urban growth and the development progress.

Urban land cover in large urban centre including metropolitan areas continually changes over time and space, and local government must be able to update their database to reflect current land use. However, conventional methods of obtaining urban land cover data require a great deal of time, effort, and money to meet fast growing cities.

Remote Sensing can provide an important source of data for urban land use/land cover mapping and environmental monitoring [264]. Urban land cover/use mapping has received an increasing amount of attention from urban planners and scientists including geographers. A number of significant studies were made for environmental quality management. Thus, various techniques have been applied for mapping urban land use/land cover. It helps in encroaching urban problems even of very small magnitude and dire.

Planning is a widely accepted way to handle complex problems of resource allocation and decision-making. It involves the use of collective intelligence and foresight to chart direction, order harmony, and make progress in public activities relating to human environment and overall development. In order to provide more effective and meaningful direction for better planning and development necessary support of the organization has become essential. Hence, the need for a suitable information system is increasingly being felt in all planning and developmental activities, whether these are for urban or rural areas. Urban areas of today are more exactly described as sprawling regions that become interconnected in a dendritic fashion (Carlson and Arthur, 2000) [265]. The positive aspects of urbanization have often been overshadowed by deterioration in the physical environment and quality of life caused by the widening gap between supply and demand for basic services and infrastructure. Urbanization is inevitable, when pressure on land is high, agriculture incomes is low, and population increases are excessive, as in the case of most developing countries of the world. Urbanization has been both one of the principal manifestation and an engine of change. The 21st century is the century of urban transition for human society. In a way, urbanization is desirable for human development. However, uncontrolled urbanization has been responsible for several problems, our cities facing today, resulting in substandard living environment, acute problems of drinking water, noise, and air pollution, disposal of waste, traffic congestion etc. To minimize these environmental degradations in and around cities, the technological development in related fields have to address to these problems caused by rapid urbanization, only then the fruits of development will percolate to the most deprived ones.

The modern technology of Remote Sensing indicated above contains both aerial and satellite data which creates opportunity to collect physical data rather easily, with speed and on repetitive basis. It offers to integrate data under Geographic Information System with reflection of analysing of the data spatially and offering possibilities of generating various options (modelling), thereby optimizing the whole planning process. Therefore, it is essential to know intensively about the characteristics and capabilities of these Remote Sensing data products available to the urban and regional planners.

6.11. Advantages of Remote Sensing and Geographic Information System Technologies in Town Planning

In India, the complexity of urban development is so dramatic that it demands immediate attention and perspective physical planning of the cities and towns [266]. The dynamic nature of urban environmental necessitates both macro and micro level analysis. Therefore, it is necessary for policy makers to integrate Remote Sensing with urban planning and management. Traditional approaches and technique designed for towns and cities may prove to be inadequate tools when dealing with metropolis. New approaches are required, and new methods must be incorporated into current practice. Until recently, maps and land survey records from 1960 to 70 were used for urban studies, but now the trend has shifted to use digital, multispectral images acquired by EOS and other sensors. The trend towards using Remotely Sensed data in urban studies began with first-generation satellite sensors such as Landsat MSS and was given impetus by a number of second-generation satellites: Landsat TM, ETM+ and SPOT. The recent advent of the third generation of very high spatial resolution (5m/pixel) satellite sensors is stimulating. The high resolution PAN and LISS III merged data may be used together effectively for urban applications. Data from IRS P-6 satellites with sensors on board especially LISS IV Mono and Multispectral (MX) with 5.8 m/pixel spatial resolution is very useful for intensive urban studies.

The output of a Remote Sensing system is usually an image representing the scene being observed [267].

Many further steps of digital image processing and modelling are required in order to extract relevant information from the image. Suitable techniques are to be adopted for a given theme, depending on the requirement of the specific problem. Since Remote Sensing may not provide all the information needed for a full-fledged assessment, many other spatial attributes from various sources are needed to be integrated with Remote Sensing data. This integration of spatial data and their combined analysis is performed through Geographic Information System technique. It is a computer-assisted system for capture, storage, retrieval, analysis and display of spatial data and non-spatial attribute data. The data can be derived from alternative sources such as survey data, geographical/topographical/aerial maps or archived data. Data can be in the form of locational data (such as latitudes/longitudes) or tabular (attribute) data. Geographic Information System techniques are playing a significant role in facilitating integration of multi-layer spatial information with statistical attribute data to arrive at alternate developmental scenarios.

Application of Remote Sensing technology may lead to innovation in the planning process in various ways:

(i) Digitization of planning base maps and various layout plans has facilitated updating of base maps wherever changes have taken place in terms of land development etc. Digital maps provide flexibility, as digital maps are scale free. Superimposition of any two digital maps, which are on two different scales, is feasible. This capability of digital maps facilitates insertion of fresh survey or modified maps into existing base maps. Similarly superimposition of revenue maps on base maps with reasonable accuracy is of great advantage compared to manually done jobs;

(ii) Since information and maps are available in digital format, correlating various layers of information about a feature from satellite imagery, planning maps and revenue maps is feasible with the help of image processing software like ERDAS Imagine, ENVI and PCI Geomatica, ILWIS. Such super imposed maps in Geographic Information System software like Map info, Arc View, Auto CAD Map and Arc GIS etc. provide valuable information for planning, implementation and management of urban areas;

(iii) Remote Sensing techniques are extremely useful for change detection analysis and selection of sites for specific facilities, such as hospital, restaurants, solid waste disposal and industry. An attempt has been made here to demonstrate the potentials of remote sensing techniques in base mapping, land use and land cover mapping, urban change detection and mapping, urban infrastructure and utilities mapping, urban population estimation, management etc.

6.12. Remote Sensing Data Access

Aerial photographs have long been employed as a tool in urban analysis. In India, city planning has been largely confined to aerial photography. It is being used for generation of base maps and other thematic maps for urban areas as it is proved cost and time effective and reliable. Wealth of information pertaining to land features, land use, built up areas, city structure, physical aspects of environment etc. are available from the aerial photography [268]. Various types of cameras and sensors black and white, colour, colour infrared are used for aerial photography. Because of security concerns related to aerial photography, the use of photogrammetric techniques was confined to smaller cities. Aerial photographs provide information that can significantly improve the effectiveness of city and town planning and management in India. They are also relatively low in cost, accurate, reliable and can be obtained on desired scale, but they are not useful in large metropolitan areas.

The overwhelming advantages of Remote Sensing methods over tactile methods based on ground survey are the consistency, which can be ensured in at least one stage of data collection, the rapidity of survey, and the small number of skilled workers required at the data collection stage: it is possible, in principle at least. Unlike the preceding data sources, the direct use of satellite imagery ensures that others have carried out a minimum of interpretation. If the imagery is being collected especially for the purpose of the survey, the client will be able to specify time and environmental conditions, which are the best for his purposes, such as midday in winter conditions. For obtaining accurate land use data the following points should be considered:

(i) *Characteristics of imagery:* scale of the photography, geometric fidelity, contrast, sharpness, resolving power of film etc., waveband(s) used, photographic or digital format;

(ii) *Characteristics of land use:* consistency of the relationship between form and function, degree of multiple use, the amount of ground control, change since imagery was obtained;

(iii) *Characteristics of user:* interpretation skill, use of stereoscopic techniques etc.

6.12.1. Town Planning Features Classification

Apart from topographical mapping, planners also look forward to Remote Sensing (data products) technology to provide them with information on existing land use and their periodic updating and monitoring. In addition, with appropriate technique and methodology the same data products can be used to:

- Study urban growth/sprawl and trend of growth;
- Updating and monitoring using repetitive coverage;
- Study of urban morphology, population estimation and other physical aspects of urban environment;
- Space use surveys in city centres;
- Slum detection, monitoring and updating;

- Study of transportation system and important aspects both in static and dynamic mode;
- Site suitability and catchments area analysis;
- Study of open/vacant space.

High spatial resolution satellite data are highly beneficial in the context of complex urban areas where relatively small size and complex spatial patterns of the component scene elements (e.g., buildings, roads and intra-urban open space) have restricted the use of the low-resolution space borne sensors. These new images increase the amount of information attainable on urban form at local level [269].

6.12.2. Town Planning Land Use Classification Requirements

Classification is, therefore, an activity of sub-dividing a group of objects in two or more groups, i.e., to arrange objects into classes according to some system or principle [270]. This can be based on activity, economic function, physical appearance, or simple land cover. The guidelines could be:

(i) The classification system should be applicable over a large area covering both city core and its surroundings;
(ii) Classification should be suitable for using Remotely Sensed data obtained at different time periods;
(iii) The minimum interpretation accuracy and reliability in the identification of land use should be about 85 percent subject to level of classification of different land uses;
(iv) The nomenclature, definition and framework to the extent possible should be compatible with existing terminologies adopted in planning agencies;
(v) Classification should be easier to understand and flexible;
(vi) Aggregation of similar or multiple land use classes should be possible at different levels of requirement;
(vii) The classes must be mutually exclusive, i.e., any geographical individual can only fall into one class;
(viii) Wherever possible, it must be based upon quantitative criteria.

6.12.3. Town Planning Land Use Delineation

For most purposes, there is a need to arrange detailed observations into groups, using some classification process. There is not a single ideal classification of land use and land cover. There are different perspectives in the classification process, and the process itself tends to be subjective, even when an objective numerical approach is used. There is, in fact, no logical reason to expect that one detailed inventory should be adequate for more than a short time, since land use and land cover patterns change in keeping with demands for natural resources [271].

Table 16. Urban Land Use Classification

Level I	Level II	Level III	Level IV
Built up	Built up (Urban)	High Density residential	High rise apartments/ flats
	Built up (Rural)	Medium Density residential	Medium rise apartments/flats
		Low Density residential	Low rise apartments/flats
			Row houses
			Slum and squatters
		Industrial	Service industry
			Light industry
			Extensive industry
			Heavy industry
			Hazardous
		Mixed built up	
		Recreational	Parks/gardens
			Playgrounds
			Historical monuments
			Cinema halls
			Swimming pools
			Major function halls
		Public and Semi Public	Educational
			Hospital
			Cantonment
			Religious
			Government institutions
			Petrol Pump
			Fire stations
			Police stations
			Others
		Communications	Post Office
			Telegraph office
			Radio TV station
			Others
		Public utilities and facilities	Water treatment plant
			Landfill/dumping
			Electric power plant
			Sewerage treatment plant
			Others
		Commercial	
		Transportation	Bus terminus
			Railway stations
			Air port
			Others
		Reclaimed/vacant land	Layouts/plotted land

6.13. LIMITATIONS OF REMOTE SENSING AND GEOGRAPHIC INFORMATION SYSTEM TECHNOLOGY USE

When Remote Sensing data is used many problems arise during interpretation of different urban land use/land cover features, in which cloud patches on satellite data is important one. Another problem in Remote Sensing data depending upon urban building size and spectral contrast with the surrounding area is that some buildings in urban area may be identified, while others may not. On the other hand, while individual houses cannot be always identified, groups of houses and city blocks can often be delineated and, in many instances, interpreted in satellite images [272].

In this context, urban Remote Sensing must be able to provide planners with certain key, data sets that are pertinent to urban studies, notably:

(i) The location and extent of urban areas;
(ii) The nature and spatial distribution of different land use categories within urban areas;
(iii) The primary transportation networks and related infrastructure;
(iv) Census related statistics and socio-economic indicators;
(v) The 3-D structure of urban areas for telecommunications (inter-visibility) and environmental Impact Assessment (EIA) studies; and
(vi) The ability to monitor changes in these features over time and space.

The following urban issues are analysed by using this technology:

Urban Land Use Inventory: It is quite natural that population growth increases the pressure on the land, and the non-urban land is converted into urban areas. Population growth and city expansion ultimately influence the land use pattern of any urban centre. Knowledge of the patterns and intensity of land use is relevant in urban planning, but the preparation of a land use inventory by conventional method is expensive and time consuming [273, 274]. The advantage of satellite imagery interpretation in terms of accuracy, timeliness, and cost is indisputable in comparison with conventional methods.

Study of Urban Sprawl and Growth Trends: Since satellite based Remote Sensing systems have unique capability to provide repetitive coverage for any pay of the world this makes it most suitable for monitoring and updating of urban expansion by using very high-resolution multi-temporal Remote Sensing data especially for town and country planning.

Space Use in the Core Area: From the monitoring point of view of city area, information about land use only may not be sufficient for city administrators and planners in congested core areas. The true picture may only be visualized from available information on actual space use. Space, however defined, has a location, and data necessarily refer to a point in time. In the absence of suitable large scale maps for such detailed studies, the principal use of very high resolution IKONOS satellite data provide a base for the survey/recording of various activities in the field. The rest were confirmed/picked up during the field visit. Quantitative determination of the space use allows the understanding of the distribution pattern of various activities and functional characteristics within urban fabric, which is useful for quantifying the stress on existing infrastructure.

Travel Route Pattern: Physical infrastructure of an urban centre comprises of transportation, water supply, electric power supply, sewage etc. The transportation system is one of the keys to rapid modernization, particularly in developing countries. Irrespective of whether roads act as a catalyst for development or play a more passive role in development, it is necessary to work out a judicious plan to avoid congestion, pollution, and cost. The main objective of this type of studies using Remote Sensing and Geographic Information System techniques is to make network analysis of tourist places considering two factors: time and cost. This type of study provides a methodology for analysing the optimum transportation network of tourist cities, and can be very helpful when planning for other cities also.

Urban Environment Analysis: Green spaces in cities exist mainly as semi-natural areas, managed parks, and gardens, supplemented by scattered vegetated pockets associated with roads and incidental locations. Embodying the garden city concept advocated by [275] and the large urban park idea expounded by Frederick Law Olmsted [276], public green spaces have been increasingly designated in cities since the 1880s to counteract environmental impacts of urban expansion and intensification. Plants notably trees, have a wide range of environmental benefits, and urban green spaces often accommodate varied assemblages of flora and small animals, providing readily accessible site with natural ingredients.

The role of Remote Sensing in the case of green / open spaces mapping and analysis has become important for managing and maintenance of old and degraded spaces. However, in a number of cases Remote Sensing can supplement or partially replace tedious ground survey methods. Moreover, ground methods having limitations, as a whole area may not be accessed in one go and information collected may not be as accurate as possible through Remote Sensing, aided by limited ground survey. Remote sensing provides not only spatial data but also allows us to compare temporal variations in spatial data, which is essential for green/open spaces management.

Site Suitability Analysis: A number of workers have identified various criteria including soils, hydrology, topography, vegetation, climate, existing built up areas, transportation route etc. to find out suitable sites for location of development activities. The most common technique for suitability analysis is weighted suitability method. Weighted suitability is more complex; in order to bias the weighting of the aspect scales they should be firstly normalized that is needed to be used in the same numerical range.

The frequently used scale is in a 1-5 range:

5. Very good (much more than average);
4. Fairly good (more than average);
4. Good (average);
2. Fairly bad (less than average);
1. Very bad (much less than average).

Such a scale can accommodate qualitative and quantitative data, but the scoring of quantitative data to such a scale needs qualified professionals. The next step is assigning the weight factors. This is, of course, the critical element in this approach. Weight factors are often based on a mixture of implicit knowledge, personal experience, and individual values those are usually called "professional judgment."

Social Infrastructure: In most Indian cities, the municipal bodies have primary responsibility to cater the basic needs of the citizen by providing required essential services and infrastructure facilities. However, during the last century, cities throughout the developing world have seen an extraordinary increase in their population size, which has put tremendous strain on the delivery of the basic infrastructure services. A major concern of municipalities in developing countries is the limited access to urban services of larger parts of the city population.

Equitable distribution is becoming the centre of concern in planning the infrastructure facilities. There is an urgent need to solve this problem of unbalanced distribution of infrastructure services. The social infrastructure facilities include banks, post-office, schools, medical facilities, etc. For each facility the proposed indicators are:

(a) Number of facility/Total population of the ward;
(b) Number of facility/Total area of the ward.

Remote Sensing and Geographic Information System is capable of extracting urban land cover information with robust results. Satellite Remote Sensing with repetitive and synoptic viewing capabilities, as well as multispectral capabilities, is a powerful tool for mapping and monitoring the ecological changes in the urban core and in the peripheral land use planning, which, will help to reduce unplanned urban sprawl and the associated loss of natural surroundings and biodiversity. On the other hand, moving further, interfacing of urban planning models with Geographic Information System should now receive due attention. Incorporation of land use transportation models, network analysis, and simulation of urban activities to evaluate different urban development alternatives in the Geographic Information System framework needs to be explored for added advantage.

Land use planning, community facilities planning, transport planning, and environmental planning all can benefit from this information. Rapid development in city poses several challenges including problems associated with urbanization for urban managers and policy makers. Meeting these challenges requires access to timely and reliable information.

6.14. Follow up Stages of Remote Sensing Method and Geographic Information System Technologies Application in Town Planning

This study assesses the extent of encroachments into the original acquisitions of Festac town using SPOT XS satellite imagery and investigates the extent of conversion of open spaces in the developed area of Festac town to other land uses using Geographic Information System techniques. Original Master plans were digitally converted and integrated with Remote Sensing and fieldwork data. The results show that encroachments into the original Festac acquisition have already been noted towards the west and north of the area. About 1.42 km^2 (142 ha) has already been affected. For the developed area of Festac, open green spaces reduced from about 0.2 km^2 (20 ha) to only 0.003 km^2 (0.3 ha). Residential houses and shops/shopping malls have virtually taken over the designated open spaces accounting for 35% and over 30% of the original areas allocated for open spaces. Only 1.6%, 4.9% and 1.4% respectively of the designated land area for open spaces still serve the purpose of functional open spaces, functional playground and open green spaces as opposed to their original allocations of 32.5%, 2.6% and 64.3% respectively. The implication of these on the liveability of Festac town, and city/urban planning is also examined [276].

Urbanization is now a common feature of all third world countries. Primate cities and megacities are emerging in developing countries. In Asia, Africa and Latin America, the unprecedented population growth that characterized much of the 20th century has evolved into unparalleled urban growth [277]. The growth in the population of cities in developing countries has been attributed to various factors, which could be summarized in two: natural population increase and rural-urban migration. The cities of the third world have especially experienced population growth. For example, the population of the municipal boundary of Lagos rose from 230,256 in 1956 to 650,000 in 1963 [278]. The 1991 census put the population of Lagos State at about 5.7m and the projected figure using a highly conservative national growth rate of 2.83% (the growth rate of Lagos is actually about 6%) is about 7.5m. Lagos is expected to add nearly 10m people between 2000 and 2015 [276]. This growth is going to be over a land mass of about 3,600 km^2 of which about 1,700 km^2 is water. Lagos has an average population density of 1,712 pp/km^2 (persons per square kilometer) in 1991. However, the bulk of the population of Lagos is concentrated on the metropolitan area – an area less than 800 km^2 - where the industrial and commercial activities are concentrated. Metropolitan Lagos has 15 of the 20 local government areas (LGAs) in Lagos State and it is a fact that some of these LGAs had population density as high as 48,000 pp/km^2 in 1991 and 65,000 pp/km^2 by the year 2000 projections.

This increased population in the city of Lagos has in turn led to widespread environmental deterioration and pollution. Formerly planned residential areas are no longer liveable. The M.J. Fasona and A.S. Omojola limit of acceptable change can be used to identify indicative capacities and requirements of an area [279]. For Lagos, evidence of development beyond the limit of acceptable change is manifested in slums and break-down of infrastructure and facilities in formerly planned residential estates. Lack of adequate tools and information for city and urban planning is also making planners in most developing countries to lose track and equally lose the battle for creating decent and liveable cities. The thrust of this study is divided into two: to assess the extent of encroachments into the original acquisitions of Festac town using SPOT XS satellite imagery; and to investigate the extent of conversion of open spaces in the developed area of Festac town to other land uses using Geographic Information System techniques.

6.14.1. Area Survey

The study area is roughly defined by UTM-31 coordinates 533409E, 713989N; 531727E, 718231N; 527512E, 718017N; and 528295E, 713512N. Located some 10 km southwest of central Lagos in-between Amuwo Odofin and Alimosho LGAs of metropolitan area, the original acquisition for Festac town covers approximately 19 km^2 (1,900ha) with a perimeter of about 220 km. Festac Town (also called Festival Town) was proposed to be a mini-town with its own facilities and employment opportunities. Therefore, the proposed area layout design of Festac town has six major land use classes: Residential, Industrial, Commercial, Recreation, circulation and central function. Like any modern residential layouts, the canals and drainage channels are well laid out to take care of flood and sewerage problems.

Festac Town, which is a Federal Government of Nigeria project through the Federal Housing Authority (FHA), was conceived around the same time that the government of Lagos State through the Lagos State Development and Property Corporation (LSDPC) hatched the idea of Planned Residential Estates to take care of the rapidly expanding population of Lagos. A portion of the residential zone covering about 5 km^2 (500ha) representing 26% of the entire planned layout area was built between 1975 and 1976 to accommodate foreign participants to the world's Festival of Black Arts and Culture – FESTAC - which was held in Lagos in 1977. Incidentally, this event gave name to the new estate, which was expanded in 1979-1980 and 1982. At completion, the estate was a model, which was the largest and first of its kind in the West Africa sub region. Up till now the remaining parts of the original layout plan including the industrial, much of the commercial, central function and much of the recreation and a larger part of the residential zones, which are yet to be developed due to the shift in emphasis to the development of Abuja (the new Federal Capital) and later, the downturn in the Nigerian economy.

6.14.2. Methodology and Data Source

This study integrates Remote Sensing data with Geographic Information System. To investigate the problem of encroachment on the original Festac town acquisition SPOT XS satellite Imagery was acquired in year 2000 and original general Land Use Zoning Plan for the area at a scale of 1:10,000 prepared in 1977 were used (Table 17).

Table 17. Data and data sources

N	Data	Scale	Year	Source
1	Land use Zoning Plan	1:10,000	1976	Federal Housing Authority
2	SPOT XS satellite Imagery		2000	Geography Department, UNILAG
3	Revised Master plan for Festival Town	1:2500	1982	Federal Housing Authority
4	Lagos State Administrative Map	1:100,000	2000	Geography Department, University of Lagos

The revised 1:2,500 Master plan for the developed area produced in 1982 was used to investigate the second problem of conversion of open spaces in already developed parts of Festac town. At the same time, it was used to generate the base data on the status of open spaces while field investigation was carried out in 2003 to see the situation on ground now. Digital file of Administrative map of Lagos State at a scale of 1:100,000 were accessed from the Department of Geography, University of Lagos.

The Land Use Zoning Plan acquired from the archives of the Federal Housing Authority (FHA) was scanned and exported into ArcView Geographic Information System. This original plan has no geographic reference, hence six control points in UTM-31 coordinates was supplied by the FHA and four of these control points were used to georeference the scanned land use zoning plan using the Image Analysis tool of ArcView. The different feature classes – land use zones and roads were extracted from the scanned image using the head-up digitizing within the ArcView Geographic Information System environment.

The SPOT XS satellite imagery covering western parts of Lagos, which had already been processed and georectified to UTM-31 coordinates was clipped with the vector layer of the land use zoning plan. Isodata classification was performed on the clipped area of the image using the Image Analyst tool of ArcView Geographic Information System. Five land use classes – Water, Built up area, Newly developing area, Forest, and Wetland were generated.

To generate data on the status of open spaces in the developed portion of the study area, the revised Master plan was scanned and imported into ArcView Geographic Information System. Garmin 12XL Handheld Global Positioning System was used to collate control points for registering the scanned map using the Image Analyst tool of ArcView. Head-up digitizing method was used to digitize and code the status of open spaces as at inception. The map generated was used as the basis for field work to check what the status of these open spaces is in 2003.

6.14.3. Data Processing and Results

An analysis of land use zoning for the whole Festac is depicted in Table 18. About 12 km^2 or 64% of the land was dedicated to proposed residential layouts. Other proposed land uses are 2.54 km^2 (13.42%) for recreational belts; 1.5 km^2 (7.95%) for light and cottage industries; and 1.44 km^2 (7.58%) for central function zone. Canals, circulation and others account for the rest.

Table 18. Land Use Zones of the Festac Acquisitions

N	ZONE	SUM_AREA (km)	Percentage
1	Canal	1.00	5.25
2	Central Function Zone	1.44	7.58
3	Circulation	0.050	0.27
4	Commercial Zone	0.26	1.37
5	Industrial Belt	1.51	7.95
6	Recreation Belt	2.55	13.42
7	Residential Zone	12.14	63.93
8	Others	0.04	0.23
TOTAL		18.99	100.00

From the SPOT XS satellite imagery, encroachments have already been recorded in some parts of the layout. About 1.42 km^2 (142 ha) has already been affected. Firstly, new developments have been noted around the already built up area. This suggests that individuals have been acquiring and developing lands on their own since the government has not carried out any development in Festac since the early 1980s. Whether these new developments are with the consent of the government through the FHA cannot be verified.

Secondly, at the western flank towards the boundary of the Festac acquisition with the International Trade Fair Complex along the Lagos-Badagry expressway, a fully developed area was noted. Field checks revealed that a Petrol Station has already been constructed here.

The third area is the northern flank with other settlements such as Egbe, Igando and Eggan. New developments have already emerged in some parts of the area that is proposed for industrial development. These new developments are scattered and are connected by tracks and untarred roads which means that the roads are probably constructed by the people themselves. These probably suggested that these developments are illegal and possibly not to the knowledge of the acquisition's supervisory authorities - the Federal Housing Authority.

The second thrust of this study is to investigate the extent of conversion of designated open places in the developed/inhabited parts of Festac. From the scanned and georeferenced revised Festival Town Master plan prepared in 1982, the designated open spaces were digitized and coded within Geographic Information System. The designation and extent of these open spaces are shown on Table 19.

Table 19. Designated Use and extent of Open spaces in Festac Town (1982)

N	Designated Use	Count	Total Area (acres)	Total Area (m^2)	Percent
1	Children playground	47	16.07	65048.33	32.45
2	Open Green Space	92	31.85	128880.73	64.30
3	Recreational Facilities	1	1.28	5175.74	2.58
4	Shop And Mechanic Workshop	1	0.26	1043.93	0.52
5	Shops	1	0.07	293.32	0.15
TOTAL		142	49.53	200442.05	100.00

From Table 19, a total of 142 spaces covering about 0.2 km^2 (20 ha) were designated as open spaces. Ninety-two of these, covering and area extent of about 0.13 km^2 (13 ha) or 64% of the entire open areas were allocated for open Green Spaces. This is followed by children playground covering 0.06 km^2 (6 ha) or 33%. Other minor designated use of open spaces includes recreation facilities, shopping complex, and automobile workshops.

a. Situation As of 2003

The map developed within the indicated implementation was taken from the field in June 2003 to investigate the scenario as in the year 2003. The results of the fieldwork are shown in Table 20. The breakdown on Table 4 clearly shows the conversion of designated open lands into other uses different from the original intended uses. Open green spaces reduced from about 0.2 km^2 (20 ha) to only 0.003 km^2 (0.3 ha). Of serious concern is that residential houses and shops/shopping malls have gradually taken over the designated open spaces accounting respectively for 35% and over 30% of the total area of conversion. It is also sad to note that only 1.6%, 4.9% and 1.4% respectively of the designated land area still serve the purpose of functional open spaces, functional playground, and open green spaces as opposed to their original allocated spaces of 32.5%, 2.6% and 64.3% respectively.

This trend may likely have resulted from increased migration of people to Festac town, but it portends dangers for aesthetics, beauty, and liveability of Festac town. It shows neglect of the city and lack of monitoring on the part of city administrators and planners. This is true more so since planners in our own part of the world still lack adequate application knowledge of the necessary tools of Geographic Information System and Remote Sensing to carry out their jobs of development control and urban monitoring.

Table 20. Appropriated Use and extent of Open Spaces in Festac Town (2003)

N	Appropriated Use	Count	Total Area (acres)	Total Area (m2)	Percentage
1	NEPA office building	1	0.9630	3,896.45	1.94
2	Abandoned	10	1.8770	7,600.48	3.79
3	Bank and mechanic workshop	1	0.5440	2,202.39	1.10
4	Bank and restaurant	1	0.9370	3,792.42	1.89
5	Church	3	1.8680	7,559.96	3.77
6	Florist	4	1.4840	6,001.96	2.99
7	Functional open space	1	0.7730	3,128.24	1.56
8	Functional playground	5	2.4370	9,865.16	4.92
9	Hospital	2	0.4660	1,886.02	0.94
10	Hotel	1	1.0050	4,068.14	2.03
11	Houses and road	1	1.6780	6,792.45	3.39
12	Mechanic workshop	1	0.4650	1,880.79	0.94
13	Mosque	1	0.5140	2,081.16	1.04
14	Office building	2	0.3130	1,267.52	0.63
15	Open green space	18	0.7090	2,876.60	1.44
16	Private garden	1	0.3840	1,554.94	0.78
17	Residential houses	35	17.4270	70,530.09	35.19
18	Restaurant	1	0.1670	675.65	0.34
19	School	1	0.2290	926.07	0.46
20	Shop	1	0.1330	537.65	0.27
21	Shopping plaza	3	1.0010	4,048.31	2.02
22	Shops	46	13.4420	54,395.86	27.14
23	Shops and mechanic workshop	1	0.3150	1,273.78	0.64
24	Shops and office building	1	0.3950	1,599.96	0.80
TOTAL		142	49.5260	200,442.05	100.00

The data and information generated in this study using Geographic Information System and Remote Sensing techniques has demonstrated the deterioration of the study area. While encroachments into the original acquisitions are discovering, there is widespread conversion of open spaces in the already developed area. The implications of this on the liveability of Festac town cannot be over-emphasized. A major problem now is that city planners and administrators in Nigeria still is lack of the required knowledge of the tools of Geographic Information System and Remote Sensing to enhance their efficiency at work. Thus, the more developed world has a duty to assist in training our city and urban planners and administrators so that we can indeed ensure liveable cities and sustainable developments on both side of the Atlantic. Only then, we can go further to bridge the Pacific and Atlantic in Geospatial information research.

6.15. An Importance of Use of Remote Sensing Method and Geographic Information System Technologies in Town Planning

Urbanization is not merely a modern process, but a rapid transformation of human social roots on a global scale, where village culture is being rapidly replaced by the urban culture. In India, the complexity of urban development is so rapid that it demands quick response and perspective physical planning of the cities and towns. Therefore, it is necessary and fundamental for policy makers to incorporate Remote Sensing into urban planning and management. Remote Sensing is the observation and measurement of objects from a distance, i.e., instruments or recorders are not in direct contact with objects under investigation [280].

Remote Sensing depends upon measuring some kind of energy that is emitted, transmitted, or reflected from an object in order to determine certain physical properties of the object. One of the most common types of Remote Sensing is photography. Till date maps and land survey records from the 1960's and 70's were used for urban studies, but now the trend has shifted to using digital, multispectral images acquired by EOS and other sensors. The output of Remote Sensing system is usually an image representing the scene being observed. Since Remote Sensing may not provide all the information needed for a full-fledged assessment, many other spatial attributes needed to be integrated with Remote Sensing data. This integration of spatial data and their combined analysis is performed through Geographic Information System technique. The fashion of using Remotely Sensed data in urban studies began with first-generation satellite.

The high resolution PAN and LISS III merged data can be used together effectively for urban applications. Remote Sensing is widely used in Cartographic surveys. P-6 data is used for updating terrain visualization, generation of a national topographic database, utilities planning and other GIS applications needed for urban areas [281]. The satellite will provide cadastral level information up to a 1:5,000 scale, and will be useful for making 2-5 meter contour map (NRSA 2005). Remote Sensing techniques are extremely useful for selection of sites for specific facilities such as hospitals, restaurants, solid waste disposal and industry. Urban planning requires a large volume of data both at the time of planning and at the time of implementation of the plan to determine the status of the available facilities. Thus, Remote Sensing techniques provide accurate, orderly and reliable information for planning and management of a town or a city. Remote Sensing techniques are extremely useful for change detection analysis and selection of sites for specific facilities such as hospitals, restaurants, solid waste disposal and industry [283].

Planning is a widely accepted way to handle complex problems of resources allocation and decision-making. It involves the use of collective intelligence and foresight to chart direction, order harmony, and make progress in public activity relating to human environment and general welfare. In order to provide more effective and meaningful direction for better planning and development necessary support of the organization has become essential. Hence the need for a suitable information system is increasingly being felt in all planning and developmental activities, whether these are for urban or rural areas. Urban areas of today are more exactly described as sprawling regions that become interconnected in a dendritic fashion [284]. The positive aspects of urbanization have often been overshadowed by deterioration in the physical environment and quality of life caused by the widening gaps between supply and demand for essential services and infrastructure. Urbanization is inevitable, when pressure on land is high, agriculture incomes are low, and population increases are excessive, as in the case of most of the developing countries of the world.

Urbanization has become not only the principal manifestation, but also an engine of change, and the 21th century, which has become the centre of urban transition for human society. In a way, urbanization is desirable for human development. However, uncontrolled urbanization has been responsible for many of the problems, our cities experiences today, resulting in substandard living environment, acute problems of drinking water, noise and air pollution, disposal of waste, traffic congestion etc. To improve these environmental degradations in and around the cities, the technological development in relevant fields has to solve these problems caused by rapid urbanization, only then the fruits of development will reach most of the deprived ones.

The modern technology of Remote Sensing, which includes both aerial and satellite based systems, allows us to collect many physical data rather easily, with speed and on repetitive basis. In addition, together with Geographic Information System it helps us to analyse the data spatially, offering possibilities of generating various options (modelling), thereby optimizing the whole planning process. These information systems also offer interpretation of physical (spatial) data with other socio-economic data, and thereby providing an important linkage in the total planning process and making it more effective and meaningful.

Recent technological advances made in domain of spatial technology cause considerable impact in planning activities. This domain of planning is of prime importance for a country like India with varied geographic patterns, cultural activities etc. [285, 286]. The purpose of using Geographic Information System is that, maps provide an added dimension to data analysis, which brings us one-step closer to visualizing the complex patterns and relationships that characterize real-world planning and policy problems. Visualization of spatial patterns also supports change analysis, which is important in monitoring of social indicators. This in turn should result in improving need assessment. The objectives of this paper are to explain Remote Sensing and Geographic Information System applications in various stages of planning, implementation, and monitoring of the urban area.

6.15.1. Approach of Town Planning Development

Urbanization or urban drift is the physical growth of urban areas because of rural migration and even suburban concentration into cities, particularly the very largest one. It is closely linked to modernization, industrialization, and the sociological process of rationalization. The rapid urbanization of the world's population over the twentieth century is described in the 2005 Revision of the UN World Urbanization Prospects report. The global proportion of urban population rose dramatically from 13% (220 million) in 1900, to 29% (732 million) in 1950, to 49% (3.2 billion) in 2005. The same report projected that the figure is likely to rise to 60% (4.9 billion) by 2030.

Urbanization has been considered as a problem in India until recently and the city governments in the country have always struggled to solve the problems. However, in recent years the perception has changed. With the cities contributing more than 65 per cent of the GDP in the recent years, the government has begun to look at the urban areas as engines of growth. With rapidly increasing population in the urban areas, urbanization has emerged as one of the most serious challenges before the planners. The first appearance of cities and urban development in India was around 2600 BC with the advent of the Indus Valley civilization.

Urbanization in India was mainly caused after independence, due to adoption of mixed system of economy by the country, which gave the rise to the development of private sector. Urbanization is taking place at a faster rate in India. Population residing in urban areas in India, according to 1901 census, was 11.4%. This count increased to 28.53% according to 2001 census, and crossing 30% as per 2011 census, standing at 31.16%. After independence, India faced poverty, unemployment, and economic backwardness. The architect of modern India, Pandit Jawaharlal Nehru, focused on the domain of science and technology, for economic development.

I. causes of urbanization in India

The main causes of urbanization in India are:

- Expansion in government services, as a result of Second World War;
- Migration of people from Pakistan after partition of India;
- The Industrial Revolution;
- Eleventh five year plan that aimed at *urbanization* for the economic development of India;
- Economic opportunities are just one reason people move into cities;
- Infrastructure facilities in the urban area;
- Growth of private sector after 1990.

II. Problems in urbanization

Lack of civic amenities is yet another problem. As per 2001-slum census, only 65.4 per cent of the households in the cities and towns had access to drinking water within their premises. Remaining households had the water supply source either outside their premises or away from their houses.

1. Urbanization - Source of lighting: Source of lighting is another important area, which was surveyed during the census. Though the percentage of households having an electric source of energy was much higher than in the rural areas, yet more than 12 per cent of the households in the urban areas did not have an electric source of lighting and had to depend on other sources like kerosene. About 0.4 per cent of the households in cities and towns had no source of lighting at all;
2. Urbanization - Availability of education facilities: In the urban areas education facilities are also key areas, considered particularly for the poor settlements. While the affluent and upper middle classes normally have the best of educational facilities available to them in the cities, the poorer sections find it hard to have access even to basic educational facilities. The level of male and female literacy rates in the slum areas is distinctly lower than non-slum population of cities, with Patna recording the highest difference of almost 30 per cent between the level of literacy rates in slum and non-slum areas of the city;

3. Urbanization – Healthcare facilities: Lack of good healthcare facilities is also an area of serious concern. As per the report of a Task Force appointed by the government of India to advice on health scenario in the urban slums, it was pointed out that 6 out of 10 children in slum areas are delivered home in Indian slums. Further, more than half of India's urban poor children are underweight and the state of under-nutrition in urban areas is worse than in the rural areas. The reach and utilization of essential preventive health services by the urban poor is generally found to be very low and about 60 per cent of the children below one-year-old are not fully immunized. Only 4 per cent couples use birth spacing methods.

In addition to the above-mentioned problems pertaining to urban and social services, there are serious gaps in the availability of infrastructure facilities in urban areas. Roads are being congested with more and more new vehicles being registered every day and parking has become a serious problem in most urban areas. Solid waste management is also a serious problem in the country, particularly in the cities. Safe disposal of the solid waste in a scientific manner is a major issue in Indian cities and towns. With over 350 million people living in urban areas and generating millions of tons of garbage every day, without proper arrangements for safe disposal of the garbage, serious problem of water contamination and environment pollution is on the anvil. The problem is worst in the areas inhabited by the poor and in the slums [287].

For the last about six decades, the government has focused on rural development and rural poverty alleviation. Billions of rupees have been spent but even after 60 years of concentration on this sector, the absolute number of rural poor in the country has actually increased. The government is now viewing the urbanization process as an alternative strategy to eradicate rural poverty. Growth rate of population in the cities of the country is much higher than the general growth rate of population in the country and there is a need to strengthen the cities and towns to be able to brace up to the challenges ahead. The government of India, in December 2005, launched an ambitious program called Jawaharlal Nehru National Urban Renewal Mission (JNNURM), for renewal of Indian cities on sustainable basis. 63 cities have been chosen under the Mission. There are also two sub-missions. First sub-mission aims at strengthening the urban infrastructure like water supply, sewerage, traffic flow, decongestion, scientific solid waste management, proper storm water drainage, preservation of heritage etc. Second sub-mission is aimed at providing basic services to the urban poor, including proper housing at affordable rates and upgradation of slums to ensure that all the slum areas in the mission cities are provided with the same level of facilities as are available to the better areas in the Mission cities.

Funding pattern under the JNNURM is that for the States of the north-east 90 per cent of the cost of projects approved under the mission is to be provided by the Union government as grant in aid and the remaining amount has to be pooled in by the Urban Local Body (ULB) concerned and the State government. For other hilly States and the cities with less than one million population, the percentage of ACA is 80 per cent. For cities with population between one million and four million, the ACA is 50 per cent of the project cost, while the same for the cities with over four million population is 35 per cent. It is expected that after including the State and ULB share, during the mission period (up to 2012-13), more than Rs 1,50,000 crore would be spent for up-gradation of urban infrastructure and for providing basic services for urban poor in the Mission cities. In addition to the JNNURM funding and projects, the city administration has to make special efforts for making sure that the challenges of urbanization are met and the cities and towns are able to cope with the urban problems in the years to come. Consequences of Urbanization Rapid rise in urban population, in India, are leading to many problems like increasing slums, decreasing in standard of living in urban areas, also causing environmental damage. The Industrial Revolution in the 18th century caused countries like United States and England to become superpower nations but the present condition is worsening. India's urban growth rate is 2.07%, which seems to be significant compared to Rwanda with 7.6%. India has around 300 million people living in metropolitan areas. This has greatly caused slum problems, with so many people over crowding cities and forcing people to live in unsafe conditions, which also includes illegal buildings. Water lines, roads, and electricity are lacking which is causing fall of living standards. It is also added to the problem of all types of pollution. Urbanization also results in a disparity in the market, owing the large demands of the growing population and the primary sector struggle copes with them.

Remote sensing is the collection of information about an object or phenomenon without making physical contact with the object. In modern usage, the term generally refers to the use of aerial sensor technologies to detect and classify objects on the Earth (both on the surface, and in the atmosphere and oceans) by means of propagated signals (e.g., electromagnetic radiation emitted from aircraft or satellites).

There are two main types of Remote Sensing:

- ***Passive Remote Sensing:*** Passive Remote Sensing means that the energy collected by the sensors is being provided by the environment. Reflected sunlight is the most common source of radiation measured by passive sensors. Examples of passive remote sensors include film photography, infrared, charge-coupled devices, and radiometers.
- ***Active Remote Sensing:*** Active Remote Sensing on the other hand, involves transmitting energy, allowing that energy to reflect off to the surrounding environment, and then recollecting that energy using a sensor. RADAR and LIDAR are examples of active Remote Sensing where the time delay between emission and return is measured, establishing the location, speed and direction of an object.

Remote Sensing makes it possible to collect data on dangerous or inaccessible areas. Remote Sensing applications include monitoring deforestation in areas such as the Amazon Basin, glacial features in Arctic and Antarctic regions, and depth sounding of coastal and ocean depths. Remote Sensing also replaces costly and slow data collection on the ground, ensuring in the process that areas or objects are not disturbed [288].

Orbital platforms collect and transmit data from different parts of the electromagnetic spectrum, which in conjunction with larger scale aerial or ground-based sensing and analysis, provides researchers with enough information to monitor trends such as El Nino and other natural long and short-term phenomena. Other uses include different areas of the Earth sciences such as natural resource management, agricultural fields, land usage and conservation, and national security and overhead, ground-based and standoff collection on border areas.

Satellite, aircraft, spacecraft, buoy create ship, and helicopter images, data were created to analyse and compare things like vegetation rates, erosion, pollution, forestry, weather, and land use. These things can be mapped, imaged, tracked and observed. The process of Remote Sensing is also helpful for city planning, archaeological investigations, military observation and geo morphological surveying

Table 21. Major regions of the Electromagnetic spectrum

Region	Name Wavelength	Comments
Gamma Ray	< 0.03 nanometres	Entirely absorbed by the Earth's atmosphere and not available for remote sensing
X-ray	0.03 to 30 nanometres	Entirely absorbed by the Earth's atmosphere and not available for remote sensing
Ultraviolet	0.03 to 0.4 micrometres	Wavelengths from 0.03 to 0.3 micrometres absorbed by ozone in the Earth's atmosphere
Photographic Ultraviolet	0.3 to 0.4 micrometres	Available for Remote Sensing the earth. Can be imaged with photographic film
Visible	0.4 to 0.7 micrometres	Available for Remote Sensing the Earth. Can be imaged with photographic film
Infrared	0.7 to 100 micrometres	Available for Remote Sensing the Earth. Can be imaged with photographic film
Reflected Infrared	0.7 to 3.0 micrometres	Available for Remote Sensing the Earth. Near Infrared 0.7 to 0.9 micrometres. Can be imaged with photographic film.
Thermal Infrared	3.0 to 14 micrometres	Available for Remote Sensing the Earth. This wave length cannot be captured with photographic film. Instead, mechanical sensors are used to image this wavelength band
Microwave or Radar	0.1 to 100 centimetres	Longer wavelengths of this band can pass through clouds, fog, and rain. Images using this band can be made with sensors that actively emit microwaves
Radio	> 100 centimetres	Not normally used for Remote Sensing the Earth

a. Stages of Urban Planning

Urban areas face many critical environmental problems, which are manifested at the time of crises. To avoid such occurrences the prime requirement is quantification, which requires a comprehensive Urban Information System (UIS) to be developed to cater the developmental needs of the growing urban areas.

Sequential steps followed by urban information system:

1. Thematic map preparation from satellite data using visual interpretation techniques;
2. Generation of spatial framework in Geographic Information System for perspective and development plans;
3. Integration of thematic maps using Geographic Information System techniques for urban sprawl analysis and urban land use change analysis.

6.15.2. Remote Sensing and Geographic Information System Applications Consequences of Town Planning

In India, the complexity of urban development is so dramatic that it demands immediate attention and perspective physical planning of the cities and towns. It is necessary and fundamental for policy makers to integrate like Remote Sensing into urban planning and management. New approaches are required, and new methods must be incorporated into current practice.

Until recently, maps and land survey records from the 1960's and 70's were used for urban studies, but now the trend has shifted to using digital, multispectral images acquired by EOS and other sensors. The trend towards using Remotely Sensed data in urban studies began with first-generation satellite sensors such as Landsat MSS and WAS given impetus by a number of second-generation satellites: Landsat TM, ETM+ and SPOT HRV. The recent advent of a third generation of very high spatial resolution (5meter/pixel) satellite sensors is stimulating.

Advancement in the technology of Remote Sensing has brought miracle in the availability of the higher and higher resolution satellite imageries. They are IRS-P6 Resourcesat imagery with 5.8 meter resolution in multispectral mode, IRS-1D Pan image with 5.8 meter resolution, Cartosat-I imagery of 2.5 meter resolution with stereo capabilities, Cartosat-II with 1 m, IKONOS imageries of Space Imaging with 4 meter in multispectral mode and 1 meter in panchromatic mode, Quickbird imagery of Digital Globe with 61 cm resolution in panchromatic mode. These high resolutions of the sensors provide a new methodology.

Apart from Cartographic applications, P-6 data will be useful in cadastral mapping and updating terrain visualization, generation of a national topographic database, utilities planning. The satellite will provide cadastral level information up to a 1:5,000 scale, and will be useful for making 2-5 meter contour map [289, 290].

The output of a Remote Sensing system is usually an image representing the scene being observed. Since Remote Sensing may not provide all the information needed for a full-fledged assessment, many other spatial attributes from various sources are needed to be integrated with Remote Sensing data. This integration of spatial data and their combined analysis is performed through Geographic Information System technique. It is a computer-assisted system for capture, storage, retrieval, analysis and display of spatial data and non-spatial attribute data. The data can be derived from alternative sources such as survey data, geographical/topographical/aerial maps or archived data. Data can be in the form of locational data (such as latitudes/longitudes) or tabular (attribute) data.

Application of Remote Sensing technology can lead to innovation in the planning process in various ways:

1. Digitization of planning base maps**:** Digitization of planning base maps has facilitated updating of base maps wherever changes have been taken place in terms of land development etc. Superimposition of any two digital maps, which are on two different scales, is feasible. Similarly superimposition of revenue maps on base maps with reasonable accuracy is great advantage compared with manually done jobs;
2. Digital format: correlating various layers of information about a feature from satellite imagery, planning maps and revenue maps is feasible with the help of image

processing software like ERDAS Imagine, ENVI and PCI Geomatica, ILWIS. Such super imposed maps in GIS software like Map info, Geomedia, Arc View, Auto CAD Map and Arc Geographic Information System provide valuable information for planning, implementing and management in urban areas.

Planning and managing cities in the new era of globalization and economic liberalization would be a demanding task calling for new skills and approach. Indian cities will have to compete with others to attract investors and, therefore, issues like quality of infrastructure, energy efficient services provision and environmental conditions in a city besides economic stability would play a significant part in such competition. Urban planning profession, in general, will have to address these issues and respond rapidly. It is worthwhile noting that spatial dynamics of cities is complex to fathom and urban theory is still static. In other words, the urban planning authorities and agencies in every parts of the country should adopt new technologies like Remote Sensing and Geographic Information System. These have capability to provide necessary physical input and intelligence for preparation of base maps, for planning proposals and act as monitoring tool during implementation phase(s). Satellite Remote Sensing with repetitive and synoptic viewing capabilities, as well as multispectral capabilities, is a powerful tool for mapping and monitoring the ecological changes in the urban core and in the peripheral land-use planning, which will help to reduce unplanned urban sprawl and the associated loss of natural surroundings and biodiversity.

On the other hand, moving further, interfacing of urban planning models with Geographic Information System should now receive due attention. Incorporation of land-use transportation models, water distribution network analysis, and stimulation of urban activities to evaluate different urban development alternatives in the Geographic Information System framework needs should be explored for added advantage.

6.16. Town Planning Development Based on Remote Sensing Data and Geographic Information System Technology Advances

Adequately planned urban development is a key to achieving sustainable environment. Urban planning therefore becomes important as we develop our urban areas. For us to maintain the land use allocation pattern/scheme of our urban areas, we need to assess, monitor and control the trend and the pattern of development as they occur as uncontrolled development (slum developments) and may mar the beauty and sustainability of the urban Area. This paper focuses on assessing and analysing of urban development using Object Based classification of Remotely Sensed satellite imageries as a part of Independence Layout, Enugu urban, Enugu State of Nigeria. Using the Object based classification, a supervised classification of urban structures and other recent physical developments in the study area were achieved. Comparative Analyses were carried out between the classified image of Independence layout Enugu and the original Town Planning design in order to determine the difference between current land use pattern of independence layout Enugu and what it was designed to be. A Quickbird image of 2012 and the original town planning design of Independence Layout were used as primary data; Erdas Imagine 9.2 was used for the object based image classification while ArcGIS 9.3 was used for the analyses. The results show the spatial, graphical (map) as well as statistical analyses of recent developments and changes in the design of the study area. The results of this study are relevant in development control by governmental authorities across the globe [291].

This paper is focused on a very vital study. In this study, urban development in Independence layout, Enugu, Nigeria was assessed using its original town planning design and an object based on supervised classified Remotely Sensed imagery of its present state of development. Urban development is the process of building up of urban areas. It requires enforcement of designs and the control of land use, which can be political and often technical. It requires researches, strategic thinking, urban and regional planning of communities, the design of a blue print for the physical development as well as an enforcement process to ensure that the development controls are observed.

A Geographic Information System is a computer aided system, which is designed, and assembled to capture, store, manipulate, analyse, manage, and present many types of geospatially referenced data. A Geographic Information System gives positional information pertaining to physical features for informed decision-making. Users apply a Geographic Information System tools and applications to create interactive queries in order to analyse spatial information [292].

Remote Sensing is the process of obtaining information or carrying out observations on an object without making any physical contact with the object. This is done using what are called sensors. The major process involved in extracting this information from the Remotely Sensed image is called image classification, in this process; we use computer based image processing software to label each pixel to particular spectral classes in order to obtain high accuracy. There are two varieties of classification procedures; they are supervised classification and unsupervised classification. The supervised classification is the essential tool used for extracting quantitative information from Remotely Sensed image data [293].

6.16.1. Area Data Collection/Methodology of Processing

Independence layout is located at a pivotal part of Enugu metropolis. It is a low-density part of Enugu urban, Enugu state, Nigeria, in Enugu North Local Government Area of the State. It is located within 335261.719E and 338025.822E and 713511.702N and 710461.335N in the UTM Zone 32 N coordinate system.

The layout is the location of the Enugu state Government as well as other key government offices such as the high Court, Appeal Court, Judicial Headquarters, etc. and it has an area of about 200 Hectares and a population of up to 120,000 residents.

The primary data used in this study were hard copy of Independence Layout town planning design (made in 1960s) obtained from the town planning the Enugu State Ministry of Lands and Urban Development, and a high resolution (0.6m resolution) Quickbird imagery of 2011 covering the study area Figure 27. The secondary data includes the ground coordinates observed for geo-referencing as well as those observed during ground trothing.

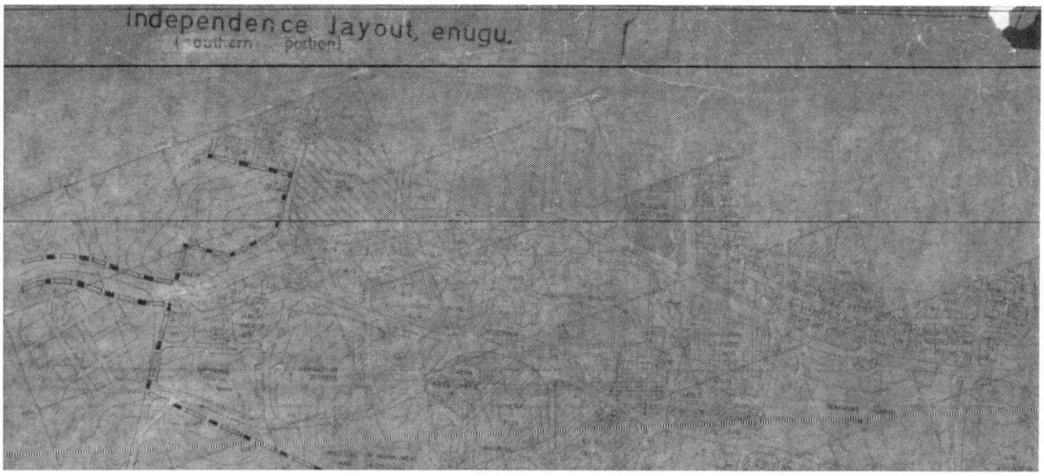

Figure 27. Data: section the Quickbird imagery of Independence Layout.

This section describes the technical processes undergone in this study. As primary data, the satellite image was already georeferenced, but the accuracy of the positions on the imagery was confirmed by comparing its coordinates observed on the ground; in order to eliminate the little discrepancies noticed, the image was further adjusted through rubber sheeting. Then it was saved in the .tiff format. The hard copy of the original town planning design of the study area, Independence layout was obtained from the town planning authorities; it was scanned into a picture format (tiff.), and was georeferenced to the same coordinate system (UTM Zone 32 N) as the satellite imagery using the observed coordinates of some ground control points and ArcGIS 9.3. After geo-referencing, shape files were created for different land uses and the design was vectorised. After the vectorization of the design, the current land use was vectorised from the satellite imagery and information about their uses was sourced from the field through. The already georeferenced image was imported into Erdas Imagine 9.2, this was then converted from .tiff format to .img format for being used in segmentation. The image segmentation was carried out on the image using user defined constraints, which controls the segmentation of different image objects into independent objects. Segmentation is the division of an image into spatially continuous, disjoint, and homogeneous regions, i.e., the objects [294]. After the segmentation, the image objects are given meaning or identification by carrying out training, which is based on the information obtained during ground trothing pertaining to the objects [295]. The image objects were diligently trained before the actual supervised image classification. In the classification process, classes were assigned to the image objects as well as the different land uses [296, 297].

During the process of ground trothing, ground coordinates of image points of the different themes (classes) were recorded as well as the different land uses. This information enabled us to check the accuracy of the image classification process that was carried out [298]. At the end of vectorization of the original town planning design process, the classified image showed the present land use pattern on the ground. At this stage, using ArcGIS 9.3, the classified (actual use) image was overlaid on the original design of the study area as vectorised from the land use design made by the Enugu state town planning Authorities.

6.16.2. Outcomes

This study was able to achieve the following results:

i. A vector map of the original town planning design of the study area;
ii. A vectorised map of the present land use pattern in the study area;
iii. A classified raster image of the study area;
iv. Statistical analyses (pie charts) showing changing in the designed land use and the actual land use;
v. Overlays and comparative analyses between the actual land use and designed land use.

6.16.3. Stage of Processing

It covers following steps of data processing:

6.16.4. Analytical Steps

In the course of the study, the following analytical steps were taken in order to assess the results listed above.

a. Analytical Step One: Statistical Analyses

These analyses evaluate the degree at which land uses and purposes vary between the design and the actual use. This analysis was carried out using the statistical analyst of ArcGIS 9.3. The pie charts were generated from the vector of the design and the vector of the actual use. See figures 5.0 and 6.0 below. If the sectors of the pies were observed diligently, it is clear that the sector for residential purposes in the design is larger than the sector for residential purposes in the actual use, this shows that the government is acquiring more land in that area because of increased government activities there. We can also observe a reduction of open spaces between the design and actual use, which means that some open spaces are being subdivided for other purposes. The pies show a great variation between the design and the actual use.

b. Analytical Step Two: Comparative Analyses between Design and Actual Use

Here, the original land use design (in vector form) of the study area was contrasted with the actual land use (classified image). In this analysis, it can be observed that many blocks have been converted to used other than the designed. For example, different land uses as can be seen in the block verged with a green line which was formerly reserved for Teaching hospital and have been subdivided into residential blocks; other places such as the General Hospital (Verged Orange), Public parking spaces (verged brown), Shopping Centre (Verged Milk colour) have been converted to government offices. Most of the open spaces have been converted to residential areas as we can see in area verged yellow (Open spaces) over brown classes (Buildings).

c. Analytical Step Three: Development Controls in the Study Area

This analysis was carried out in order to assess the degree of development control in the study area. If the Vector/Raster overlay in figure 8.0 below is observed, it will be noticed that the road network design has been altered largely in particular, in the southern part of the study area. From the classified raster, it will be noticed that the road network is narrower than the originally designed road width; this indicates that developers encroach into public lands and roads. This may be because of little or no development control; the problem stern from the institutional organization of urban planning in Nigeria and such leads uncontrolled development [299].

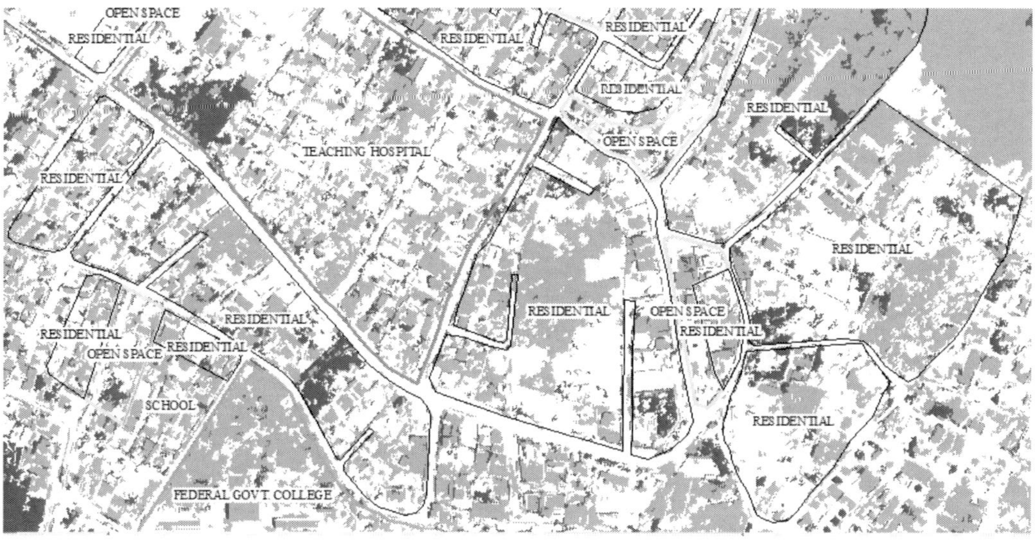

Figure 28. Developments in the study Area.

In conclusion, this study is timely and relevantly considering the present land reform issues in Nigeria. The study has highlighted the benefits of assessing urban development using supervised classification of remotely sensed imageries and Geographic Information System. The methodology demonstrated in this study can be applied by the town planning and lands officers in development control, urban planning, land allocation, re-allocation, etc. This method employs a quick and cost effective way that underscores its advantages over the traditional methods of town planning and development control being applied in the study area currently.

So many observations have been made in this study and salient issues were noted. The research team deemed it necessary to discuss the following points as regards the application of this study:

(i) Urban development in Nigeria should be assessed from time to time and Town planning designs should be updated from time to time using the methods demonstrated in this study. This is because, it was observed that there was no update in the planning design of independence layout scheme since the first design was made;

(ii) Proper measures should be put in place by the appropriate Government Authorities in order to enforce development controls, as little or no development controls were observed in the study area;

(iii) The Government and communities in undeveloped areas instead of mutilating or further subdividing the already planned areas should plan new layouts. These practices may turn planned cities into slums;

(iv) Surveyors and geo-informers should insist on government approved town planning designs/schemes before carrying out any cadastral or layout survey for any client;

(v) It was observed that the problems of urban planning in Nigeria stern from the Land tenure system; that is what differentiates the implementation of the Land Use Act in the Northern Nigeria and that of the Southern Nigeria. Land tenure system remains a barrier to the institutional context of urban development planning and control in Nigeria.

6.17. Space Technology Advances Application in Town Planning

The possibility of use of Remote Sensing methods and Geographic Information System technology capabilities for town planning is described in this book. The main target is to demonstrate of successful application of Remote Sensing and Geographic Information System technology in data collection, analysing and processing which beneficiary provide impact for management and decision making in town planning [300].

6.17.1. Requirements of Use

There is no doubt that more employment opportunities exist in high density, large population centres which leads and needs to the relocation of people to new cities for discharging of the situation. Intensive examination of a new urban area requires comprehensive understanding of the physical, political, social, and economic dynamics. For this purpose, Remote Sensing technologies offer that they can efficiently and objectively quantify the physical characteristics and growth of cities [301-304]. There are following aims and main tasks required in developments:

(i) city area selection;
(ii) vertical layout;
(iii) underground utilities;
(iv) landscaping;
(v) social facilities;
(vi) natural and artificial reservoirs improvement;
(vii) urban areas lighting;
(viii) health improvement.

It is obvious that natural condition and factors are also important in town planning. It has to be considered natural factors in selection of areas of settlement for town planning. The main natural factors have been reflected below:

(i) relief of selected territory has to meet the requirements of site development, urban transport and drainage system achieving by gravity approach;
(ii) territory should not be selected on the waterlogged or water flooded areas;
(iii) soil resistance has to meet of intended type of site development;
(iv) residential areas should be located in the windward side against of the air pollution sources and upstream in case of the river presence against of industrial facilities provoking pollution of the water reservoir;
(v) territory needed to be maintained with the best, satisfied sources of water supply and suitable areas for sewage system;
(vi) territory should be considered for a further expansion;
(vii) territory needed to maintain external highways and railways or to be allowed of those facilitation with a low cost expenses in case of absence;
(ix) settlement territory should not be content of minerals having industrial value.

Other important territory selection circumstances in town planning:

(i) climatic condition of selected area/for full climate characteristics are:
   - monthly average temperature;
   - relative air humidity;
(ii) land cover/land use of the area;
(iii) engineering - geological conditions.

6.17.2. Remote Sensing Methods and Geographic Information System Technology in Town Planning

Remote Sensing can provide an important source of data for town planning mainly as an excellent instrument on land use/land cover mapping and environmental monitoring. This technology makes available to collect required information for huge number of features of town planning. Town planning land cover/use mapping has received an increasing interests from town planners and specialists involved in indicated implementations.

Planning is a widely accepted way to handle complex problems of town planning for further decision-making by appropriate state authorities [305]. The modern technology of Remote Sensing which includes both aerial and satellite based systems allows us to collect physical data rather easily with speed and on repetitive basis. Together with Geographic Information System, it helps to analyse the data spatially, offering possibilities of generating various options (modelling). Thereby it allows optimization of the whole planning process. These information systems also offer interpretation of physical data with other socio-economic data and thereby providing an important linkage in the total planning process and making it more effective and meaningful. Therefore, it is essential to know intensively about the characteristics and capabilities of these Remote Sensing data products available to the urban and regional planners.

6.17.3. Main Requirements for Town Planning

Apart from topographical mapping, planners also look forward to Remote Sensing (data products) technology to provide them with the information on existing land use and their periodic updating and monitoring. In addition, with appropriate technique and methodology the same data products can be used to:

(i) Study possible growth/sprawl and trend of growth;
(ii) Updating and monitoring using repetitive coverage;
(iii) Study of expected town morphology, population estimation and other physical aspects of town environment;
(iv) Space use surveys in high population and facilities density city centres;
(v) Study of transportation system and important aspects both in static and dynamic mode;
(vi) Site suitability and catchments area analysis;
(vii) Study of open/vacant areas.

High spatial resolution satellite data is highly beneficial in the context of complex town planning areas where relatively small size and complex spatial patterns of the component scene elements (e.g., buildings, roads and town empty space) have restricted the use of the low-resolution space borne sensors. These new images thus increase the amount of information supporting to manage effectively town planning process with detailed consideration expected features in town-selected area.

.6.17.4. Land Use Classification Criteria for Town Planning

Classification is an activity of sub-dividing a group of objects in two or more groups, i.e., to arrange objects into classes according to some system or principle. This can be based on activity, economic function, physical appearance, or simple land cover. The guidelines could be:

(i) classification system should be applicable over a large area covering both city core and its surroundings;
(ii) classification should be suitable for using Remotely Sensed data obtained at different time periods;
(iii) minimum interpretation accuracy and reliability in the identification of land use should be about 85 percent subject to level of classification of different land uses;
(iv) nomenclature, definition and framework to the extent possible should be compatible with existing terminologies adopted in planning agencies;
(v) classification should be easier to understand and flexible;
(vi) aggregation of similar or multiple land use classes should be possible at different levels of requirement;
(vii) classes must be mutually exclusive, i.e., any geographical individual can only fall into one class;
(viii) wherever possible, it must be based upon quantitative criteria.

6.17.5. Conceptual Approach of Space Technology Use in Town Planning

It is important to investigate areas for a new town planning necessary for meeting all the main requirements of engineering facilities development, social, natural and human aspects. It has been studied conceptually the areas embracing the territory of Azerbaijan and offered a way of a new town planning in case of necessity of execution such implementation.

Figure 29 reflects the view of an area for a new town planning. In this regard, space images have been processed with content of foregoing mentioned stages demanded to be undertaken in a new town planning execution process. It has been sectioned of the area into six zones: residential, down town, and areas of green, school, kindergarten and parking.

Figure 29. (Continued).

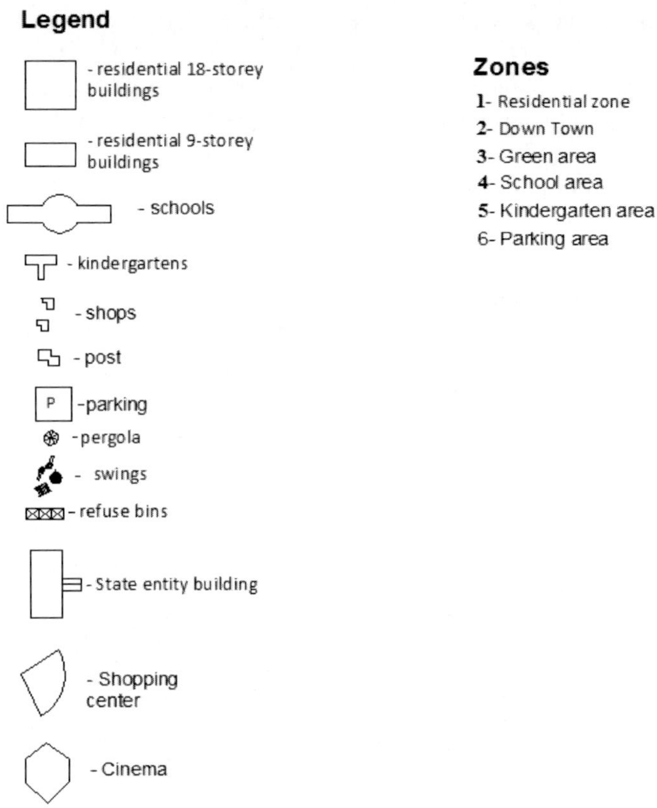

Figure 29. Town planning integrated into the space image/topographic map.

The use of data collected by means of Remote Sensing methods with further Geographic Information System technology application opens a good environment for a suitable location of all engineering and social structures, which are highly important aspects of the successful town planning development. The legend shows the main features of the city necessary for satisfaction of life needs. In the other hand application of space technology outcomes makes available, easy integration of city engineering and communication systems to the existing in outside of the city maintaining services.

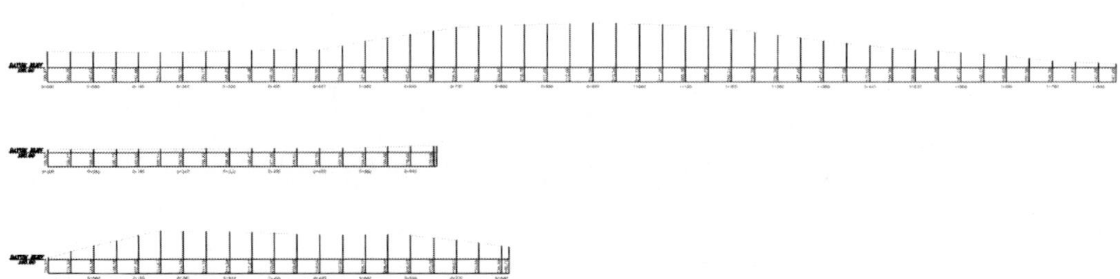

Figure 30. Elevation data of selected area for town planning.

One of the important issues is the data related to the evaluation model of the selected area for the town planning. Space image data processing with orientation of Digital Evaluation Model (DEM) is the way of effective understanding and management of industrial location and required other important technological options of town planning process (Figure 30). It is an excellent advantage for human safety aspects with forecasting of any possible natural disaster impacts as a whole and minimization its consequences.

There is no doubt that purposeful approach of engineering execution comes from successful achieved data collection related to all areas of developments. There is several methods can be used for this target. The modern technology applications particularly space technology advances with Remote Sensing and Geographic Information System technology use become a good instrument in challenge of expected outcomes of town planning. It relates a wide aspect of town planning as a good choice of starting stage of town planning as well as detailed management of the issues as a whole embraced full stages of execution.

The paper demonstrates conceptual approach of the application of the space technology in town planning process.

CONCLUSION

Results presented in this chapter demonstrate the advantage of use of the space data in town planning. It is important to indicate that a success of use such a technology depends on how successfully data accessing and processing is necessary to select.

In the other hand, filed data and measurements are affecting the quality of outcomes. It is highly important how to use and based on what Remote Sensing needed to be selected for processing. It will have an impact of the development of Geographic Information System.

In this regard, it can be adequately explained stages of space technology application in town planning.

Chapter 7

ARCHITECTURAL DESIGN WITH USE OF SPACE TECHNOLOGY ACHIEVEMENTS

ABSTRACT

Remote Sensing method and Geographic Information System most actively are used in architectural design. It is very sensitive area of engineering developments where advance technology required to be applied in design process. For this reason, it is highly important to develop conceptual approaches of architectural design. At the same time, the selection of methodology, requirements and standards are demanding for design and construction processes. The main issue is to integrate architectural design segments and filed data into the system which could operate as one system.

In this Chapter conceptual approach of space technology application in architectural design has been considered. It has been found in different areas of engineering applications related to the area of architectural design circumstances.

Keywords: space technology application, architecture, design, construction, engineering solutions, engineering achievements

INTRODUCTION

An architectural design demands the variety of conditions and requirements in implementation. It takes a number of stages needed to be undertaken during all stages of executions. It is very important when applying the space technology achievements in architectural design processes is considered. For this reason, below segments are necessary to involve into the execution framework:

- Initial data any available to access;
- Area selection for implementation;
- Area data collection;
- Space data;
- Space data processing;
- Integration space and existing data;

- Results;
- Decision making.

This indicated stages reflect the main architectural design process. It is obvious that there are a number of elements did not include the content of architectural design.

7.1. Remote Sensing and Geographical Information System Applications in Architectural Design

An architectural design is embracing primarily design of buildings or urban landscapes, as opposed to the construction documents and management required for its construction. For successful implementation of design visions, it is necessary to have a good creative skills, imagination and artistic talent in design. Modern requirements of architectural design demand an ability of integration of a number of engineering disciplines. It relates especially in regeneration elements of design particularly in urbanization [306].

Figure 31. The sketch of Murtuza Mukhtarov Street with the transport net system.

It has been suggested the regeneration of one of the central streets of Baku Murtuza Mukhtarov street, the capital of Azerbaijan, Figure 31. There is a proposal to take down mainly one-and two-story houses, which do not have architectural significance. Instead of indicated circumstances the houses or shops that are falling into the architectural ensemble are to be constructed. In some areas between the buildings to pave the squares is suggested.

There is no doubt that existing complexity of the design from the point of consideration all required features of the area it is necessary to use and apply the recent advances of modern technologies. It has been demonstrated a success of implementation of architectural design with an approach of space technology methods. Remote Sensing was used as a good instrument for data collection with detailed consideration of the area and around of features such as existing buildings (date of construction, number of buildings story, type of buildings-offices and leaving houses etc.), communication system, environmental and ecological condition of the area, statistical information. As the consequences, those collected data were integrated into the geographic coordinate system with further Geographical Information System development. It is obvious that the success of the design will depend on the number of data integrated into the Geographical Information System. It will allow considering comprehensive approach during the development of design and construction processes. In the meantime, the use of such technology creates an excellent environment for the project management as a whole. At the same time, an issue of impact of possible natural disaster and hazardous consequences for risk assessment and scale of damages has been recommended. The use of data based on the processing space images was demonstrated for the purpose of architectural design. The advantages of the use of the space technology and their weaknesses for the applied case are reflected. The result of investigation attempt was applied for the street regeneration design of "Murtuza Mukhtarov" which highly needs for renovation and restoration undertaking the requirements of design.

7.2. Remote Sensing Methods and Geographical Information System Technology in Design Process

Today, more than at any time before, buildings and other structures are incorporating a multitude of new technologies, materials and processes. There are several reasons for this, including:

- environmental and ecological issues;
- safety aspects in view of increasing natural hazards;
- attractiveness in terms of a building being a corporate 'icon'; and
- Growing pride of homeowners in their properties.

7.2.1. Design and Concepts

Architects, designers, builders, environmentalists and – last but not least – consumers have already begun to embrace new technologies in the areas that promise lower energy consumption and hence lower running costs.

7.2.2. New Requirements

Discussions with the insurance and re-insurance sector show that they have ever-greater concerns about the statistical increase in 'extreme weather' in recent years, coupled with the fact that more and more dwellings are being built in earthquake and flood-prone zones. In the wake of the devastating earthquake in Turkey in 1999, the idea of attempting to design an 'absolutely earthquake-safe building' was born. It is therefore proposed to adapt European building standards to these trends and to enable structures to cope with:

- wind speeds of up to 220 km/h (10-second gusts);
- flooding to depths of up to 3 m;
- earthquakes of up to 7.5 on the Richter scale;
- subsidence of 1.5 m during the lifetime of the building;
- severe hail and exceptionally heavy rain, and
- For some areas of southern Europe, bush-fire resistance.

7.2.3. How Space Technology Helps?

Discussions with city officials indicate that there is an urgent need to upgrade many inner-city areas to make them much more attractive places for living and working. In addition, some city authorities would like to see new buildings constructed in such a Building Sector.

Based on the various discussions outlined above, and a review of recent and expected future research and technology development activities in the space sector, efforts within the ESA Technology Transfer Program (TTP) have been focused on three major areas of application for the available space technologies:

- safe houses/buildings with respect to the natural and human-induced environment;
- healthy houses/buildings with respect to medical and mental well-being;
- Ecological houses/buildings with respect to sustainable energy supply, natural resource consumption and manufacturing processes.

The objective of the work has been undertaken for identification of regeneration of the central street of the city, based on the collected data for the further development of Geographic Information System using Remote Sensing methods. Studied area of the city has been discovered using the field observation guided by existing streets information and application of Global Positioning System technology. The spatially referenced database is satisfied with the need for accurate and easily accessible street features information. Field crew has been collected Global Positioning System point data at the street embracing all features as a whole within the investigated area. The Global Positioning System gathered data during the fieldwork were entered into the Geographical Information System. The Geographical Information System was used to generate statistical, spatial and thematic data. A comprehensive set of spatially referenced streets zones information has been generated (Figure 32).

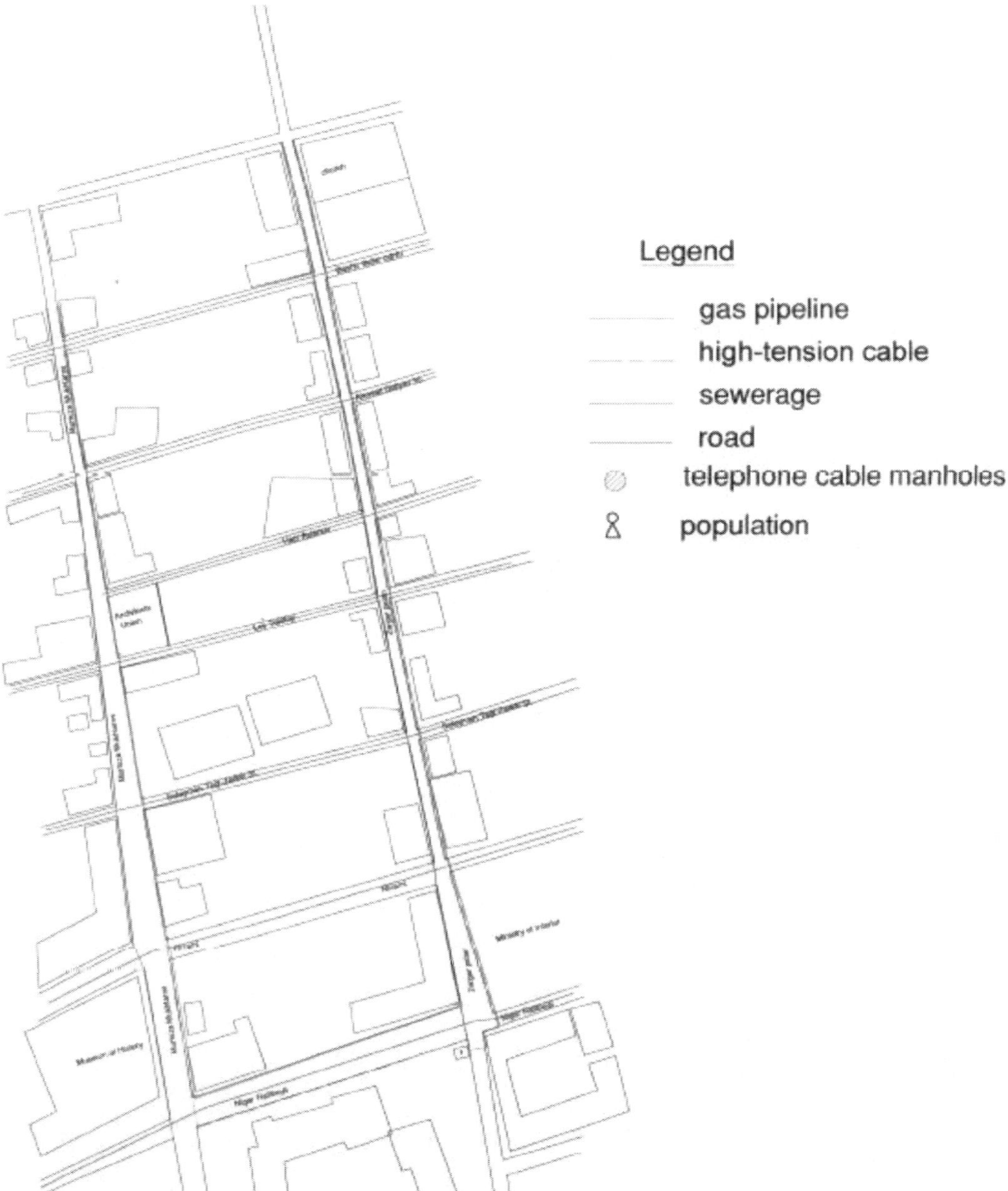

Figure 32. The features of the selected area measured by the Global Positioning System.

The steps of research achievements consist of following:

- Research – all pertinent and relevant existing data at the streets in the studied area were collected and reviewed;
- Field Investigation – field crew using GPS units, collected latitude and longitude readings at the pipeline crossing of public roads, railways and rivers within the studied area;
- Geographical Information System Development – Global Positioning System data collected by the field crew has been used for development of pipeline system layers based on Geographical Information System;
- Outcomes – Geographical Information System has been developed with the reflection of generation of the statistical, spatial, and thematic data. In the meantime, hard-copy maps and digital Geographical Information System files were produced. The final product of the investigations was the unique database specifications as the digital base maps layers.

The resulting spatially referenced database should satisfy the need for accurate and easily accessible information for regeneration of the street Murtuza Mukhtarov. The database can be important tool for the protection and utilization of the significant resources.

Due to the existed maps of the streets with inaccurate and/or too generalized information, there is a need to develop the Geographical Information System at least for some parts of the selected area. There is also a need for uniformity, since the maps contain varying levels of accuracy and scale. A Geographical Information System street mapping has many advantages over viewing spatial data on a map. A Geographical Information System mapping allows a level of analysis such as establishing networks creating buffers and modelling distance decay, which is not possible to produce on the hard copy maps. This approach is successfully applicable for linear system applications [307, 308].

It is a highly important to have a sufficient information related to the streets location for the reason of consideration of any available elements needed to be enclosed as a part of regeneration process. In case of any information, imperfection can be dangerous for implementation of successful regeneration of the street from the point of view of architectural design and social issues. The street systems have crossing elements. There is a huge communication system integrated and operating as a whole system. It demands the application of high accuracy modern technology advances for being able to consider all available data and management capability.

The need for the development of the Geographical Information System for street regeneration is also related to population protection in the selected area. On the studied streets along selected, Murtuza Mukhtarova and Zargar Palan are populated around 1220 people (Figure 33).

Knowledge of the statuses of the streets is the significant issue for many entities, including city and region authorities, emergency officials, business and industry representatives. This study provides the relevant and updated information being satisfied with the needs of customers responsible for street regeneration and staff who has involved in an architectural design of the city.

Figure 33. Space image data processing for selected street undertaken for regeneration.

7.2.4. Methodology

At the first stage of investigation, it was necessary to collect existing information along the streets for further understanding and clarification of the status. This information has been undertaken for Geographical Information System development of street regeneration.

In the next stage, the maps were used for identification of the existing configurations in the investigated area. Based on this the present status of the area and field investigation protocol were developed.

7.2.5. Field Methodology

The field components of the study as the presence of buildings, crossings, communication systems (high-tension cables, gas pipeline, sewerage, and telephone cable manholes) have been involved. It has been usually used the presence of a "witness post" that provides evidence for the field investigations. Pipelines often follow "rights-of-way," long avenues of clearing where no development occurs. The streets are required to be marked by "witness posts" during processing. For researchers involved in street regeneration the "witness post" contains the same useful information as the operator name and an emergency contact phone number.

Field staff was experienced in the use of Global Positioning System and Geographical Information System which were employed to collect required data among selected area. These field crews were directed to collect Global Positioning System point data at the streets with all crossings and other features.

7.2.6. Geographical Information System Development

The further stage after the Global Positioning System data, digital photographs and other documentation were collected based on each field investigations and transferred into Microsoft Excel spreadsheet. A method was applied for data projection and conversion and subsequent Geographical Information System development.

7.2.7. Data Projection and Conversion

The deliverable of the digital data for this project must be seamlessly integrated with the existing data. This stage of action is required. It is necessary to merge all digital data generated for this project based on a Gauss Kruger coordinate system for Zone 8 using Pulkovo 1942 as the reference datum related for the investigated area. Therefore, all data brought into the project Geographical Information System were converted to Gauss Kruger.

ESRI's ArcGIS/ ArcInfo 9 suite of Geographical Information System software were used throughout the work. The resulting deliverables in projects are in the form of ESRI .shp files. ArcGIS/ArcInfo is vector-based software that is the industry-standard for Geographical Information System - based research. ArcGIS software includes the following suite of integrated applications: ArcMap, ArcCatalog, and ArcToolbox as well as, 3-D Analyst, Spatial Analyst, Network Analyst, Geostatistical Analyst and needed for work other extensions. Each of these applications was valuable for conducting the Geographical Information System research for this work. ArcMap, a map-centric application was used for a basic mapping and editing tasks, as well as analysis. ArcCatalog was used to manage geographic and tabular data, create, and edit metadata. ArcToolbox was used for data conversion and geoprocessing. ArcInfo, an enhancement to the standard ArcGIS software was also used because it provides advanced editing capabilities [309, 310].

A base Geographical Information System was developed using the following data sources:

.shp files of digital vector data:

- 1: 100 000 scale topographic maps (PULKOVO 1942 datum);
- Mosaicked 12 LANDSAT Enhanced Thematic Mapper Plus (ETM+) scenes;

- Digital orthophoto generated from QuickBird or IKONOS imagery for most sensitive zones in the study area;
- For more detailed analysis of the streets, the orthophotos are generated from very high spatial resolution satellite images like IKONOS (1 m).

The coordinate values in the latitude and longitude columns of the .dbf file of GPS points representing the location of streets were displayed as x and y values using ArcMap. These values were exported as a .shp file and their projection defined as latitude/longitude (PULKOVO 1942) using ArcToolbox. Although these Global Positioning System points were stored in the latitude/longitude reference system, the file displays properly when added to the project GIS for the reason of projection "on-the-fly" to Zone 8, PULKOVO 1942.

In the stage of the final editing, the process was completed and the vector layers in the .shp files representing different data has to be further processed in order to create topology. Topology is the procedure for defining spatial relationships, such as line connectivity, in a database.

It might consider following approaches based on above indicated options:

1. It has been suggested the regeneration of the Murtuza Mukhtarov street in Baku, Azerbaijani capital undertaking the use of modern requirements of the architectural design challenges;
2. Due to the high density of human, social, technological and communication features of the selected area for regeneration, the use of technology was demanded which makes available containing all information sources internally;
3. The space technology capacity was used and applied for the purpose of street regeneration as a successful tool for these implementations.

7.2.8. Conceptual Approach of Space Technology Applications in Architectural Design

It has dedicated for the space technology advances use in architectural design for selected area of Gobustan in Azerbaijan. It describes conceptual approach of the process of the use of space technology in initial stages of design process [311].

The indicated technology for collection of necessary data during execution and decision making of engineering solutions has been applied. An advantage of use of the space technology is an excellent instrument of merging and integration accessed data into geographical coordination stimulating easily those definition and importance of all stages of the engineering executions.

Application of technological advances in architectural design became very important in project execution. It relates not only to quality of project implementation but also it embraces the best demonstration of expected design, timely delivery of all aspects of contenting design issues, convenient source management – human, financial etc.

There is no doubt that it has to be forced by appropriate methods, technical and technological fundaments that can be applied in all stages of design and construction processes. They can be reflected as follow:

- Method definition;
- Required Technology Capabilities identification and definition;
- Preparation works – staff and required equipment, including software;
- Advance technology application.

It is vital issues to consider area features in architectural design in particular in town planning follow up executions. What is needed to be undertaken in the process of expected architectural design in the stage of area selection for town planning? We are expressing our vision of conceptual approach of use of the space technology advances in architectural design.

7.2.9. Space Technology – Area Selection Circumstance

Location of Gobustan is in the south-eastern spur of the Great Caucasian Range and is situated on 60 km south of Baku, capital of Azerbaijan. It is a monticules semi-desert area dissected by numerous gullies and ravines. The region contains caves and outcropping rocks. The area settled since the Stone Age is one of the major and most ancient museums of rock engravings (petroglyphs) in the World.

In Gobustan – in the area of this fantastic destruction of mountains such as Beyukdagh, Kichikdagh, Jingirdag, Shongardag and Shikhgaya, the 'sea of rocks', the witnesses to the past of Azerbaijan people of the Stone Age and subsequent periods are concentrated: these are rock carvings, settlements, tombstones etc.

7.2.10. Why Gobustan?

In 1966, Gobustan was designated as a National Protected Heritage Site with an area of 4400 ha. The main aim of the open-area museum is to preserve the Gobustan site while keeping it open to the public. It makes possible to have a visit the site today and Gobustan now attracts thousands travellers and tourists interested in the early history of civilization from all over the World.

People as well as scientists are interested in this area for a wide reason. Gobustan Rock Art Cultural Landscape covers three areas of a plateau of rocky boulders rising out of the semi-desert of central Azerbaijan, with an outstanding collection of more than 6,000 rock engravings bearing testimony to 40,000 years of rock art. The site also features the remains of inhabited caves, settlements and burials, all reflecting an intensive human use by the inhabitants of the area during the wet period that followed the last Ice Age, from the Upper Paleolithic to the Middle Ages. The site, which covers an area of 537 ha, is a part of the larger protected Gobustan Reservation.

Gobustan has outstanding universal value for the quality and density of its rock art engravings, for the substantial evidence of the collection of rock art images presenting hunting, fauna, flora, and lifestyles in pre-historic times and for the cultural continuity between prehistoric and mediaeval times that the site reflects [312].

It creates an excellent environment for the development of required facilities for visitors of the area for their accommodation and arrangements. This circumstance demands to design the further construction of buildings and appropriate infrastructure in the area. In conformity with this, it is necessary to consider large aspects of architectural design and construction standards and requirements embracing all stages of engineering activities.

7.2.11. Pre - Architectural Design Data of the Selected Area

Undoubtedly, it has to be done a number of executions before architectural design open advantages for the success of the intended project implementation. It can be presented as follow:

- Historical value of the selected area;
- Geological requirements and condition (soil, relief etc.);
- Communication (water, power and gas supply, sewerage system, phone/internet/TV; system, other technical facilities);
- Buffer zone (detailed studies);
- Ecological and environmental studies.

a. Environment

The climate of Gobustan is a dry semi subtropical one, with rather mild winters and very hot dry summers. An atmospheric precipitation is small and so are the springs of Big Caucasus area. The presence of rare and wild flora testifies that the climate of Gobustan some tens of thousands of years ago was damper and the atmospheric precipitation was sufficient, so springs were abounding with water.

In the remote past, the flora and fauna of Gobustan were incomparably richer. Its landscape represented as a kind of savannah with corresponding climate. There were large herds of wild bulls, goats, deer, wild horses, donkeys, wild boars, and gazelles hunted by lions, wolves, wild cats, and leopards.

The vegetative world of Gobustan has a character that is common for deserts and semi-deserts. It consists of ephemeris grasses and bushes, wormwood and similar long-term plants. Among heaps of stones and rocks a wild rose, a dwarfish cherry, Hibernian honeysuckle, a juniper, wild pear, wild fig, wild pomegranate, grapes and some other kinds of trees and bushes are rather often met decorating the stern landscape.

b. Fauna

The fauna of Gobustan has strongly grown poor for the last decades of years. The natural inhabitants of Gobustan now are rare foxes, jackals, wolves, hares and wild cats, mountain chickens, wild pigeons, larks alongside with numerous snakes and lizards and some others.

c. Mud Volcanic

Azerbaijan and its Caspian coastline are home to nearly 400 mud volcanoes, more than half of the total throughout the world. In 2001, one mud volcano 15 kilometres from Baku made world headlines when it suddenly started ejecting flames 15 meters high.

On the average, every twenty years or so, a mud volcano may explode with great force in Gobustan, shooting flames hundreds of meters into the sky, and depositing tons of mud on the surrounding area.

d. Seismic Data

It has been studied and investigated the following areas from the seismic point of view of Gobustan:

Tectonic and sediment transport overview;
Regional Cross-sections;
Seismic Profile Interpretation;
Stratigraphic Columns;
Reservoir Characterization Petrology/Petrography;
Formation Waters;
Clay Data;
Mineralogy of Reservoirs.

Although a relationship between the occurrence of large earthquakes and the eruptions of close mud volcanoes is well known, several uncertainties remain on understanding the triggering mechanisms. The effects of two earthquakes of M w 6.18 and 6.08 occurred in the Caspian Sea on 25 November 2000 close to Baku city, Azerbaijan. A total of 33 eruptions occurred at 24 mud volcanoes within a maximum distance of 108 km from the epicentres in the 5 years following the earthquakes.

The overall eruption rate in the studied area of the 50 years before the 2000 earthquakes was 1.24 that is much smaller than the eruption rate of 6.6 of the 5 years following these earthquakes. The largest number of eruptions occurred within 2 years from the earthquakes with the highest frequency within 6 months. The calculation of the earthquake-induced static effects shows that crustal dilatation might have triggered only seven eruptions at a maximum distance of about 60 km from the epicentres and within 3 years. Based on this data, dynamic rather than static strain is likely to have been the dominating "promoting" factor because it affected all the studied unrest volcanoes and its magnitude was much larger.

7.2.12. Space Technology in Architectural Design

Gobustan area of Azerbaijan has been selected for town planning. For studying the areas stereo topographical survey was used.

a. Stereo Topographical Survey

This method is the method of development of the original topographical map based on the processing of the photographical images of the area by means of stereo topographic or photogrammetric. The method as a result of stereo topographical survey makes possible an interpretation of aerospace images, conducts stereoscopy acting of relief and develop of original maps.

The aim of the stereo topographical survey is the stereoscopy vision that is an ability of human eyes to sense the three-dimensional space [313-316]. The three-dimensional model is understood as decreased spatial optical model of the area defining in case of crossing of aero or space images generating of stereoscopy pairs. Processing aero or space images (relief survey and horizontal application) are implementing the use of stereo photogrammetric facilities.

Figure 34 reflects the map based on the method described above. Figures 34-36 are space images used for selected area with a different unglues and views integrated to the stereo topographic data.

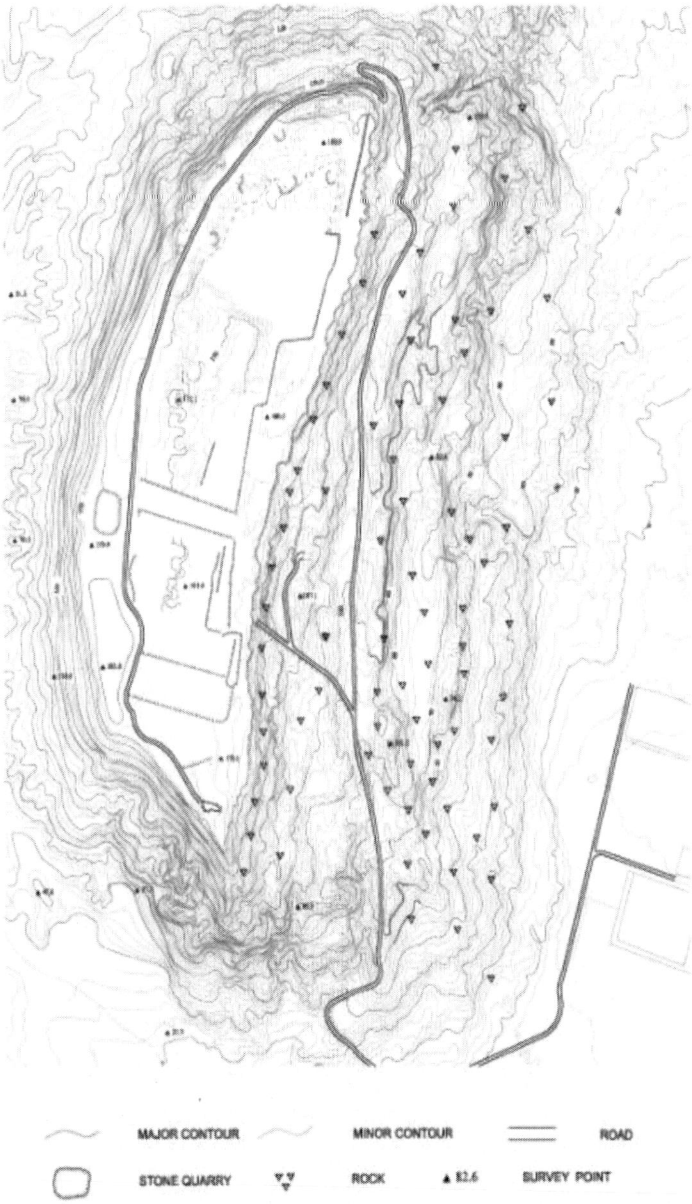

Figure 34. Stereo topographical Survey.

Figure 35. Top view of the selected area.

Figure 36. A view of the area with orientation South-North.

Figure 37. A view of the area with orientation North-South.

This describes conceptual approach of the use of the space technology advances for town planning process. Gobustan area of Azerbaijan for architectural design of the town has been selected. It reflects stages of the design process before, during and as-built part project execution. There are circumstances of Remote Sensing methods and GIS technology applications creating positive environment for purposeful and successful use of collected data for design execution.

It is intended to propose actual architectural design in the selected area with the use of the space technology in the next step of investigations.

7.3. Architectural Design of Yacht

7.3.1. The Yacht Tradition and Culture in Azerbaijan - General View

The Azerbaijan Republic is the biggest country in the Caucasus region located at the crossroads of Western Asia and Eastern Europe [317]. It is bounded by the Caspian Sea to the East, Russia to the North, Georgia to the Northwest, Armenia to the West and Iran to the South. The exclave of Nakhchivan is bounded by Armenia to the North and East, Iran to the South and West, while having a short borderline with Turkey to the northwest.

Baku is the capital and largest city of Azerbaijan, as well as the largest city on the Caspian Sea and of the Caucasus region. It is located on the southern shore of the Absheron Peninsula, which projects into the Caspian Sea. The city consists of two principal parts: the downtown and the old Inner City. Baku's urban population at the beginning of 2009 was estimated at just over two million people [318]. Officially, about 25 percent of all inhabitants of the country live in the metropolitan city area of Baku.

The city is the scientific, cultural and industrial centre of Azerbaijan. Many sizeable Azerbaijani institutions have their headquarters there, including SOCAR (State Oil Company of Azerbaijan Republic), onc of the world's top 100 companies and others [317]. The Baku International Sea Trade Port, sheltered by the islands of the Baku Archipelago to the east and the Absheron Peninsula to the north, is capable of handling two million tons of general and dry bulk cargoes per year [319]. Baku hosted the 57th Eurovision Song Contest in 2012.

7.3.2. Transport Capacity

Baku's Central Railway Station is the terminus for national and international rail links to the city. The Kars–Tbilisi–Baku railway, which will directly connect Turkey, Georgia and Azerbaijan, began to be constructed in 2007 and was scheduled for completion by 2012 [320]. The completed branch will connect Baku with Tbilisi in Georgia, and from there trains will continue to Akhalkalaki, and Kars in Turkey [321]. Sea transport is vital for Baku, as the city is practically surrounded by the Caspian Sea to the East. Shipping services operate regularly from Baku across the Caspian Sea to Turkmenbashi (formerly Krasnovodsk) in Turkmenistan and to Bandar Anzali and Bandar Nowshar in Iran [322]. The commuter ferries, along with the high-speed catamaran Seabus (Deniz Avtobusu), also form the main connection between the city and the Absheron peninsula [323].

The Baku Port was founded in 1902 and since then has been the largest Caspian Sea port. It has six facilities: the main cargo terminal, the container terminal, the ferry terminal, the oil terminal, the passenger terminal and the port fleet terminal. The port's throughput capacity reaches 15 million tons of liquid bulk and up to 10 million tons of dry cargoes [324]. Beginning in 2010, the Baku International Sea Trade Port is being reconstructed. The construction will take in three stages and will be completed by 2016. The estimated costs are 400 Million US$ [325]. From April to November the Baku Port is accessible to ships loading cargoes for direct voyages from Western European and Mediterranean ports.

The State Road M-1 and the European route E60 are the two main motorway connections between Europe and Azerbaijan. The motorway network around Baku is well developed and is constantly being extended.

The International Airport is the only commercial airport serving Baku. The new Baku Cargo Terminal was officially opened in March 2005. It was constructed to be a major cargo hub in the CIS countries and is actually now one of the biggest and most technically advanced in the region [326]. There are also several smaller military airbases near Baku, such as Baku Kala Air Base, intended for private aircraft, helicopters and charters [327].

From the I century AD till the VII century Baku bay was dry and the islands in the bay joined the land. In Ptolemy's map Baku was described far from the sea. After the VII century began the rising of water level of the Caspian Sea until the IX century and since then formation of Baku bay began [328]. Severe changes were happened at the end of the XIII century, when the Caspian Sea raised up to more than ten meters. Mario Sanuto, Italian geographer of the XIV century sadly noted that: "Water level of the Caspian Sea rises each year and most cities have flooded," and according to words of Abd ar-Rashid al-Bakuvi, the geographer, in 1403 the Caspian Sea flooded the part of Baku and water stood near a mosque. At that time, the sea was waving near foot of the Maiden Tower, which confirms a legend about a girl, who threw herself from the top of the tower. At that time Sabayil Castle, which was built on the rock in Baku bay, was totally located under water. During the next 600 years, level of the sea was hesitatingly stood high until the beginning of the XX century, when it began to sink.

Figure 38. A view to the bay from Bayil hill.

7.3.3. Environmental Condition

According to a statement of a representative of the Ministry of Ecology and Natural Resources, there are a lot of reasons of pollution of Baku bay (Figure 38). Wastes from industrial, dwelling and domestic objects located on the seashore and also exploitation of vessels, oil and gas deposits are the main factors that worsen the ecological condition. Works on investigation of depth of Baku bay began after 1996, and later works on extraction of metallic constructions and sunk vessels from water were began. In 2007-2009's, large scale works on cleansing of the bay from sunk vessels, metallic constructions and non-exploited hydro technical constructions, non-serviceable subsea pipelines began [329]. In whole 4500 tons of scrap, metal and approximately 412 tons of wooden wastes were extracted from the sea [330].

Figure 39. Baku Yacht Club.

Trade port is stretched along the northern and western shores of Baku bay as it is shown in Figure 39 and 40. The port is fully re-equipped; all loading and unloading works in it are mechanized. There is a port for ferry quay in the north-western part of Baku bay. There is a discussion on the need of construction of a bridge through Baku bay, - from Lokbatan to Zikh highway. It is considered that length of the bridge over the sea, from Shikh to Zig is 14.5 km. Total length of the bridge, which will be connected with encircling highway is 26 km. The sea bridge will be constructed on high piers, for not impeding movement of vessels. It is planned to deliver the bridge until 2015. Automobile highway through Baku bay will be paid. Construction works are undertaken for last 3.5 years.

Figure 40. Part of the bay. A view from funicular.

Foreign companies plan to open a recreational centre in Boyuk Zira and Pirallahi islands of Baku bay. In Azerbaijani newspapers is reported that a Turkish company which is preparing an appropriate project will hold the construction of the recreational centre in Nargin Island. It is planned to open modern hotels, entertainment places and other buildings in the island. It is also planned to open one of the biggest Luna parks in the CIS, which will be constructed by German specialists. It is expected that the island will be one of the centres of tourism in the future.

7.3.4. Caspian Sea

One of the interesting features of the Caspian Sea is the change of its level with sharp falling and lifting of various duration. The reasons of fluctuations are not established until now, though it is supposed, that it can be connected with fluctuation of a drain of the rivers, bottom deformation, with receipt of waters under the Earth, climate changes. Geological and paleo geographical researches allow glancing the remote past of region. For the last of ten thousand years the level of the Caspian Sea, located below level of world Ocean, varied in limits from minus 20 until minus 40 meters of region. The level of the sea in IV-II centuries BC stood on a mark not above minus 36 meters, in VI century it was on an absolutely mark minus 34 meters, in X century minus 29 meters, in the beginning of XIV century minus 19 meters, and now it is nearby minus 27 meters.

As measurements showed, for the first century (until 1930) the sea level did not change almost, fluctuated between marks minus 25 and minus 26 meters. During 1929-1941 there was a sharp fall almost on two meters (from minus 25.88 to minus 27.84 meters). Further for next thirty-seven years the sea increasing going down in the level, slowly fell, in 1977 during measurements reached the lowest point minus 29.01 meters. In connection with the sea, deviation in these years has been expanded Baku embankment parkway and the bottom terrace is equipped. Then the sea started to raise quickly and had risen on 2.35 meters in 1995 and reached a point minus 26.66 meters. If water that running into the Caspian sea from the rivers didn't understand volume about forty cubic kilometres in a year for economic needs and an irrigation (it corresponds to ten centimetres in a sea level), this level would be on one and a half meter above, coming nearer to record-breaking high for hundred sixty years to marks of the last century. The next years the sea average level had decreased almost on 50 cm, reached a point minus 27.17 meters in 2001. Since 2001 and on present time level began to raise again: in 2002 on 2 cm, in 2003 on 4 cm, in 2004 on 8 cm, in 2005 on 12 cm. At present, the level of the Caspian Sea has reached -27.0 meters. Results of studying of atmospheric processes and global climatic changes give the chance to assume that the sea level will raise on 1 m in 2015 from a present absolute mark.

Change of the level of the Caspian Sea in a current XX century is resulted in the following Figure 41.

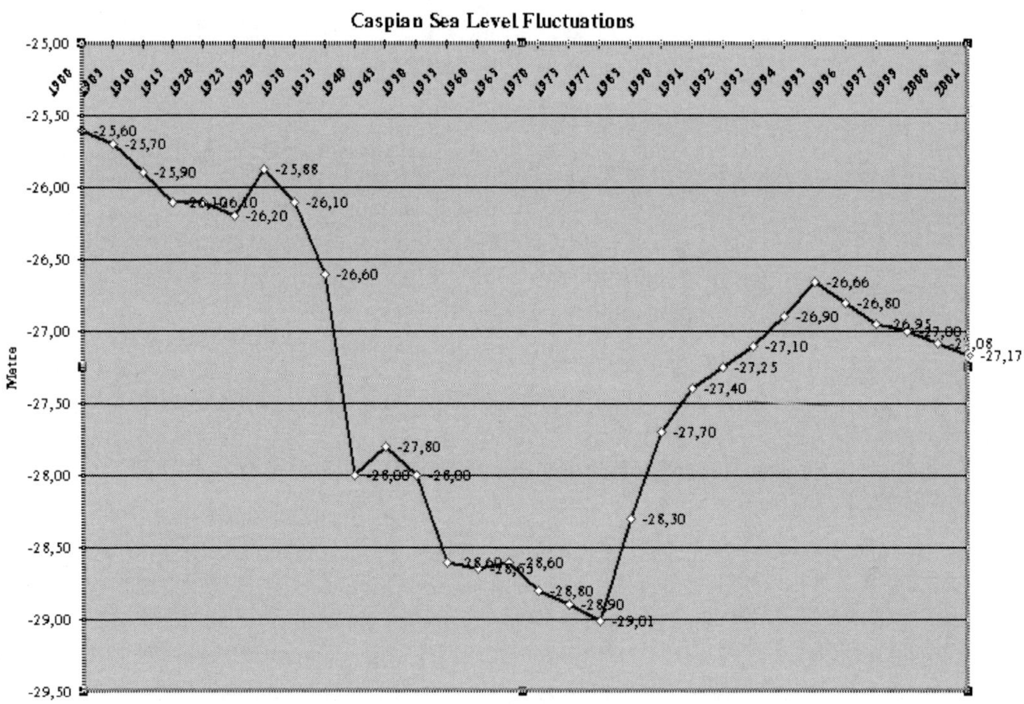

Figure 41. Caspian Sea coastal line change.

Sea level change influences on economic activities of people and on the nature. In consequence of sea fall there is a swallowing of approaches to ports, navigation conditions become complicated, the mode reserves, and stocks of fish changes. Level increase leads to flooding of coastal territories on which there are settlements of people, agricultural grounds, and the industrial enterprises.

In the Baku bay the sunk and semi sunk vessels, which have left operation sea chisel installations, moorings, and also interchisel metal ways are dismantled and transported on coast. These works are carried out according to "The comprehensive plan of actions for improvement of an ecological situation in the Azerbaijan Republic for 2006-2010." The Baku bay is already cleared of scrap metal. On a site of oil and gas extraction management "Qum adasi" oil platforms unsuitable to operation, the interchisel metal ways, unused chisel and other hydraulic engineering constructions also are cut and transported on coast for secondary use.

Historians interpret an origin of the name of a city differently. Some explain that the word "Baku" occurred from "Bud-Cuba"-blow of winds, or "Bud-Kiu" a city of winds, others consider that the city received the name from a tribe "bakan" or "bugs" which occupied Absheron in XII-V centuries BC.

The arms of Baku-torches are burning against a background of sea waves. In fact, Azerbaijan is called "the country of fires" and Baku with its natural fires-some kind of its symbol. The arms look like shield-shaped plate of rectangular form, the bottom party in the form of a sharp corner. The plate from different directions is framed by a golden strip. Background of arms is light blue colour (a symbol of the world and clear sky) and against a plate, three tongues of flame identical size (a symbol of the Country of Fires) golden colour are represented. In the bottom part of the arms the site of black colour, the sea waves of turquoise colour occupying the rest is represented 1/10 part of a plate occupying on a vertical line and showing presence of oil.

Baku has a thousand-year history. Materials of archaeological excavations (1932-1936) which have shown to argue that into place of Baku there was a settlement. It is also known, that in VIII century inhabitants of these places extracted oil.

In the XII century, Baku for a short time becomes the capital of Shirvanshakh. Erection of a fortification around old city concerns this period. By XII century value of Baku increases and Baku bay gets big popularity as good port.

Baku is the largest traffic centre of Republic. Here, navigable, railway, automobile and aviation means of communication cross. The convenient geographical position of Baku on seacoast provides the ample opportunities for transportation of its production, and first of all mineral oil. The Baku port is pivotal of country ports.

Trading port will extend along northern and western coast of the Baku bay. The port is completely converted now, all loading and discharging in it are mechanized. In a northern part of Baku, there is a mooring for a ferry. Iron and highways connect Baku with other cities of Republic and large cities, neighbouring countries.

7.3.5. Yacht Wharf Construction

a. Requirements for Selection

One of the important issue yacht wharf construction is determining vision, desire and needs in the first step of design process. It is necessary to have an approach for comprehensive problem solution based on a wide vision and consideration of all possible matters during the construction. Various options should be discussed for floating plans and fixed pier plans, including boat dock berthing needs, dock construction materials, and an overall boat dock layout. With this conceptual plan in mind, it can be moved forward to evaluate specific site selection.

b. Site Plan Evaluation

The success of construction depends on how to direct the process. For this reason, it is necessary to study selected for construction area during the period of low water to evaluate the feasibility of proposed yacht wharf vision. In the meantime, the wave exposure, bottom substrate, tidal wetlands, environmental resources, and other site conditions, which may influence the feasibility of proposed yacht wharf design, should be evaluated. With a preliminary yacht wharf plan and layout in mind, a waterfront survey can be initiated.

- The wharf concept design does not propose any coastal engineering structures, which will influence coastal processes. The wharf design is adapting to the existing wharf alignment and therefore it is unlikely that there will be an impact on the marine ecology due to an impact on coastal processes;
- The concept design proposes to extend the existing wharf using a pile construction. The impact on the sea grass is therefore limited to the locations of the new piles only, and therefore it is limiting the impact on the seabed;
- An Environmental Management Plan (EMP) should be developed to manage the construction phase of development to minimize the impact on marine ecology. The EMP should state that construction activities must avoid the Weedy Sea dragon breeding season; and
- The wharf concept design proposes to utilize the existing old wharf piles where they meet engineering requirements. Therefore, a proportion of the marine pile community will be maintained.

c. Dock Design and Dock Plans

The yacht wharf design and planning stage is the most important for the success of the project. To initiate this phase is necessary for evaluation of all of the critical data gathered to this point and weighing the numerous variables to begin formulating yacht wharf plan. The yacht wharf planning and design stages should undertake vision and desires for being the most important criteria in this process. However, there is a requirement to evaluate other variables including water depth, tide range, bottom substrate, wave exposure, environmental resources, and the regulatory permit restrictions of your state. At this point, personal requests should be considered for yacht wharf construction materials whether it is traditional wood, concrete, aluminium, synthetic materials, or exotic hardwoods. Using this information other details can be formulated like evaluate and prepare the best yacht wharf design and yacht wharf plans to meet specific yacht wharf needs - making yacht wharf vision a reality.

For the successful implementation of the project on yacht wharf, the Master Plan development is the key instrument. The Master Plan is a concept design for the pier that has been developed to respond to the following elements:

- User needs;
- Engineering requirements;
- Budget constraints;
- Marine ecological considerations;
- Heritage considerations;
- Landscape and visual impacts;

- Coastal processes; and
- Integration with the yacht wharfs design and Management Plan.

The general condition of the yacht wharf construction has been described in several engineering reviews as deteriorated. Therefore, the concept design has assumed that reasonable re-work may be required. The following is a summary of the concept design proposed:

- A vehicle turning ability consideration;
- A location for a future crane needed to propose for the central service landing area;
- Installation of a crane is however the subject to commercial viability, and would be required to be operated by a commercial entity;
- The location of the lower landings of timber decking construction needed to be maintained and in case of necessity extended. During detailed design, an alternative material such as fibre-reinforced polymer should be considered for the lower landings.
- Disabled access needed to be undertaken.

The following specialist studies are undertaking to provide existing condition information and its surrounds:

- Marine Ecology;
- Cultural Heritage;
- Landscape and Visual Assessment;
- Geology and Geomorphology; and
- Coastal Processes.

The above studies needed to be summarized.

In conformity with this for project success satisfaction, the following primary mission needed to be considered:

- System approach and develop solutions that are both economically and environmentally feasible;
- Personalize services to meet the project needs and budget;
- Cultivate satisfied yacht wharf design and plans through the development of successful projects.

The concept design of wharf responds to the heritage considerations and opportunities as following:

- The concept design for the wharf is utilizing the existing wharf alignment and materials where possible;
- The concept design is maintaining the existing wharf; and
- An underwater archaeological survey will be undertaken to identify and mitigate potential impacts on heritage values.

The concept design of the wharf responds to the landscape and visual considerations and opportunities:

- The concept design of the wharf is maintaining the length, configuration and form of the wharf and although the wharf is proposing to be wider for the first section up to the central service landing area, the visual appearance, in particular, the view from the available place will not change significantly;
- The wharf designing to utilize the same materials where possible;
- The wharf design is maintaining the existing wharf alignment, consideration is giving to rotating the lower landings to be parallel with the predominant waves however following consultation with users this is not considered as necessary;
- The design has increased the surface area of the pier thereby making more area available for use by recreational and commercial users, including the lower landing tie up areas;
- The wharf design provides separation between vehicles and pedestrians, making it safer;
- The wharf design adapts to the existing pier alignment and therefore improves access for all existing wharf users;
- Provision of a turn-around bay at the mid-point of the pier improves access for commercial users; and
- The wharf design allows disabled access to the upper section of the wharf.

The objective of waterfront survey is to obtain an accurate and detailed mapping of specific waterfront area. A detailed survey will be translated into a more accurate and detailed floating wharf design and plan or fixed wharf plan in the later project phases.

Following the waterfront site surveys, all upland and in water survey points are downloaded right from the survey data recorder to our in-office computers. At this point, experienced appropriate drafting and design team can use state-of-the-art computer aided design (CAD) software to map the details of the waterfront site. The precision of this technology and mapping process allows the best possible dock design in the final project phase.

d. Developed Drawing Applications

The yacht wharf design and plan application drawings are one of the most important components to the success of yacht wharf permit and subsequent yacht wharf design and plans project. To initiate this phase it is necessary to evaluate all of the critical data gathered to this point and weigh the numerous variables to begin formulating yacht wharf design and plans application drawings. The vision and desires is considered to be the most important criteria in this process. However, other required data is desirable to be undertaken including water depth, tide range, bottom substrate, wave exposure, environmental resources, and the regulatory permit restrictions of the state. Using this information design stage can be started and evaluated with further development of the best yacht wharf design, plans application drawings to meet specific needs and standards for making yacht wharf design, and plans vision in implementation stage.

e. Regulatory Permit Implementations

Considering the submittal of the yacht wharf design and plans permit applications to the regulatory agencies and they typically begin their permit review process. During this process, these agencies will often have questions regarding the proposed project plans. In this stage of the construction, providing the responses to any state entity questions in an effort to expedite their review and achieve final permit approval is required.

f. Construction Management Process

The next stage of the yacht construction is to start of a yacht wharf design and plans for construction and management phase.

For instance, it is understood that aquaculture operators require the use of up to 10-ton vehicles on the wharf on a semi – regular basis. It can be considered during the concept development process that vehicle access will be on a limited basis (approximately 10 per day). Therefore, by having a single turning point at the central service landing area, there should be sufficient time between vehicle movements to allow full access for the aquaculture operators.

In this case, the proposed concept allows for 10-ton vehicles, which should allow for:

- Aquaculture operators;
- Refuelling Vehicles; and
- Emergency Vehicles (excluding heavy fire trucks).

The concept design for turning area has considered the use of a crane. In detailed design, an area of operation for the crane will be specified to allow the safe use. An ideal location for a crane to operate is in the central service landing area. This area has a higher concentration of wharf and hence will have a higher capacity for accepting the larger loads experienced by an operating crane. During detailed design, the actual crane capacity and the exact location for operation will need to be fully developed.

Disabled access is available to the upper section (excluding lower landings) of the entire pier. It may be possible to allow disabled access to commercial vessels from the mid-point of the pier. To have disabled access provided to the lower landings would require a significant increase in the wharf footprint in order to allow 1 in 14 ramp to each of the landings.

Construction staging is considered in the development of the concept. Due to the lower than expected final cost for the project it is considered more economical conditions in order to construct the entire project in a single stage.

However, in order to gain a more accurate cost estimate, the following stages should be undertaken:

- Perform a full as-built review including a condition assessment of the wharf piles at the seabed. This appears to have not been performed before. This assessment will allow a greater understanding of what remedial actions would be required to the wharf;
- Perform a detailed geotechnical investigation including drilling of the area proposed for additional wharf. By having a better understanding of the geotechnical characteristics of the wharf foundations the detailed design can be more accurately

and efficiently designed and the likelihood of construction issues due to foundations will be significantly reduced;

- Develop a detailed design. Following the detailed design the budget contingency can be significantly reduced as the final arrangement is understood to a much more detailed level;
- Perform remedial works to areas that will remain in the final arrangement; and
- Undertake an underwater Archaeological Survey to identify potential impacts on heritage values.

Prior to construction of the wharf, a Cultural Heritage Management Plan and an Environmental Management Plan may need to be prepared and various approvals obtained from Commonwealth, State and Local Government agencies.

The above investigations and works could be considered in the first stage of the project and would require less mobilization than the major stage of the project.

If the construction phase can be divided into several stages, there is likelihood that the overall costs will increase, as there will need to be multiple mobilizations and the risks associated with the costs estimation will also increase.

7.3.6. Construction and Management Works

There is no doubts that such a project demands big investigations. For this reason, it is required to make sure that everything is done correctly during all stages of yacht wharf design and plans including construction. It can be evidenced by arranging project implementation supervision.

A construction part of the project is a part for project implementation. The yacht wharf design and plans going to construction phase which is the most exciting part of the project – going from paper plans to implementation. It is necessary to develop comprehensive approach of yacht wharf design and plans for construction and management execution, which is the main issue of the success of project completion.

7.3.7. Cost Specifications

It is necessary to mention that the bid specification package for yacht wharf construction is important. The objective here is to specify the exact yacht wharf design and plans components and hardware for assurance of a quality yacht wharf design and plans construction project, as well as prepare a competitive bid package to be able to obtain fair pricing from various yacht wharf design and plans.

7.3.8. Completed Project Inspection and As-Built Surveys

As a part of any construction work, it is necessary to assure that yacht wharf design and plans as well as construction have been done in compliance with the conditions of regulatory yacht wharf design and plans, match the yacht wharf design, plans, and follow good engineering practices. Upon completion of the yacht wharf as-built survey of the completed yacht wharf design and plans should be done to meet the requirements of the regulatory permits. It contains a part of Contract signed between Parties.

7.3.9. Space Technology Advances Application

Remote sensing is the measurement of object properties on the Earth's surface using data acquired from aircraft and satellites. It attempts to measure something at a distance, rather than in situ, and, for this research's purposes, displays those measurements over a two-dimensional spatial grid, i.e., images.

Remote-sensing systems, particularly those deployed on satellites, provide a repetitive and consistent view of the Earth facilitating the ability to monitor the Earth system and the effects of human activities on the Earth. The Earth's atmosphere absorbs many electromagnetic wavelength band-length ranges. The electromagnetic wavelength band ranges transmittable through the Earth's atmosphere are sometimes referred to as atmospheric windows.

As it was mentioned previously the human eye only detects visible wave range of the reflective solar radiance that human actually sees and is the part of the electromagnetic wavelength scale in the band length range 0.4 – 0.7 μm. But Remote Sensing technology allows the detection of other reflective and radiant (e.g., thermal) energy band-length ranges that reach or are emitted by the Earth's surface, and even some Earth's atmosphere reflects, e.g., the electromagnetic wavelength reflective qualities of clouds. Hence, for viewing purposes red, green, and blue (RGB) false colour assignments are used to express the reflective qualities of objects in these electromagnetic wavelength band-length groups, and the combination and mixing of these false colour assignments express the true physical reflective qualities of all objects present in an image.

The primary benefit of Global Information Systems is the ability to interrelate spatially multiple types of information assembled from a range of sources. These data do not necessarily have to be visual. Shape files are helpful for interpolating and visualizing many other types of data, e.g., demographic data. Many study and research models rely on the ability to analyse and extract information from images by using a variety of computer available research tools and then express these findings as a part of a project with images in a variety of layers and scenes.

When utilizing satellite images to assess most types of land cover change, primarily those involving change in vegetation coverage, variations in climate must be considered. For better control and accuracy in these analyses, comparing images acquired during the same month or season is advisable. However, due to the limited availability of satellite images, obtaining materials corresponding both spatially and temporally to the location and period under research are not always possible.

Furthermore, annual and seasonal climate data are not always available for the region or temporal period being researched. Sometimes, the changes in average rainfall, temperature, etc. must be inferred using more macro regional or global data.

One standard Remote Sensing application for detecting temporal change in land cover, especially vegetation, is the Normalized Difference Vegetation Index (NDVI). The NDVI application involves a ratio formula between the visual red and NIR EM bands. This ratio application helps to distinguish healthy and stronger vegetation reflection from other materials with similar reflective qualities in those EM band wavelength groups. NDVI applications are useful because two images can be processed into a false colour composite, which allows visual temporal change detection in vegetation coverage. Moreover, by applying standardized thresholds to multiple NDVI manipulated images, one can create classification training regions and execute supervised computer-generated classifications of multiple images. From these resulting images, area summary reports are calculated. These empirical data enable a more accurate assessment of change in an area of the corresponding land-cover classes. Information pertaining to some of the above topics, as well as a more comprehensive description on some Remote Sensing technologies including a glossary of terms, is given in the sections below.

Figure 42. The map of the selected area for wharf construction.

This technology can be successfully used for the wharf construction. This technology is an excellent instrument for data collection as well as in the area of construction and engineering. It allows using management process with more efficiency and low cost in point of the construction stage and for the period of future.

It was used an example wharf place in Baku, Absheron peninsula, Azerbaijan. As a first stage, initially most suitable place was selected for this purpose, considering all available data (maps, filed data, including meteodata, space and airborne images, etc.) which needed to be integrated into the Geographic Information System which is the best way for the access in any needed case.

Figure 42 reflects location of the wharf constructed a few years ago. It is important to note that the meteodata is highly important in this matter thank to a reach and strong wind behaviour of the place. It opens an opportunity to create facilities for a sailing vessel kind of services, which is an additional advantage of the place.

Figure 43. The wharf in Baku, at the Absheron peninsula, Azerbaijan.

Figure 43 demonstrates the wharf in Baku based on Remote Sensing data processing. The technology based on space technology application can be updated easily in case of any available features marked on site and surrounded area.

CONCLUSION

This Chapter reflects conceptual approach of the use of the space technology advances, particularly Remote Sensing and Geographic Information System in architectural design. There is an architectural design example for the selected area and demonstration of Remote Sensing and Geographic Information System in successful implementation of expected engineering solutions.

It has been demonstrated a wide possibilities of applications of the space technology advances in other areas of architectural design for capture of needs in engineering.

Chapter 8

Environmental Aspects of Space Technology Contribution

Abstract

One of the negative consequences in engineering is a negative impact on the environment. For instance, our intention is to generate electric power. Because of the process, we pollute the air and warm our rivers or oceans. Our goal is to extract oil from the ground. However, developing so we spill some of it, polluting a fragile landscape or seashore. There is intend to build a highway to allow people to move easily from one place to another.

The time has come how to solve the problem, how to stop or minimize the consequences? There are two ways for engineers to affect this situation. The first is to clean up existing environmental degradation. The second approach is design the way, how to minimize the degradation. For this reason, it is necessary to collect appropriate data and process for further identification of the reasons and consequences of environmental problems.

Keywords: Environment, Remote Sensing, area selection, engineering activities, space image and data processing, analysis and decision making

Introduction

There are discussions and debates about engineers needing to be educated for the environment. Engineers are not only expected to have some form of environmental education but also to take an active role in helping to solve environmental problems. This idea has come about because engineers seem as the cause of problems but also as the solution to environmental problems.

There is no doubt that the many of the environmental problems of the present have been caused by technical developments for which the engineering profession has been at least in part responsible. On the other hand, many of the solutions to the same problems are also technical and, again are the responsibility of the engineer. So, it is very important to consider as much as possible affecting reasons and consequences in engineering environment.

8.1. An Environmental Impact Assessment with Use of Space Technology Advances

Achievements in information systems, satellites imaging systems and improved software technologies have led to opportunities for a new level of information products from Remote Sensed data. The integration of these new products into existing response systems can provide a wide range of analysis tools and information products on the base of developed Geographic Information System. Using the higher resolution of space imagery and change detection analysis natural disaster awareness and damage assessment can be conducted rapidly and accurately [331].

The assessment of environmental conditions subjected to oil pollution of soil and water in some selected areas of Absheron peninsula of Baku, Azerbaijan is offered. Soil pollution was graded into five conditions from low value of oil contamination to very high.

At the beginning, stage field data was collected and reflected in the map by grades indicated above. The polluted soil sections were coordinated and integrated with the space image used for further processing. The method has been applied for a wide area impacted by oil contamination that allowed assessment of environmental conditions in the selected area.

The methodology proposed for assessment can be successfully used for permanent monitoring of situation for further dynamic change of environment.

Oil industry in Azerbaijan is one of the most profitable areas. It was known from manuscripts of most Persian and Arabic that oil production in Absheron peninsula started in the Middle Ages. The oil industry itself started here in XIX century when one the most known oilfields called Bibi Heybat was place in operation. Multiple oil and gas fields are being explored nowadays. Today oil industry in Azerbaijan is a leading economical trade.

Unfortunately, human activity changes the nature. Preservation of the environment was not the main point for people in XIX and early XX centuries. Oil production was conducted without any measures taken to reduce the damage to the environment. As a result we have some territories on Absheron peninsula looking like it is presented in Figure 44 below.

Figure 44. Typical polluted oilfield at Absheron peninsula.

Oil and petroleum derivatives affect natural processes and their interdependence. It influences not only soil, which becomes useless for agrarian and construction needs, but also air, water, vegetation and fauna.

Taking into account the above mentioned points, we understand that it is highly important to develop ranking of different types of Absheron soils, from highly contaminated to non-contaminated. This problem could be solved with the use of large volume of environmental monitoring data for separation of territories carried out during long-time period. Unfortunately, environmental monitoring in Azerbaijan is a new practice and we do not possess such volume of information. Obtaining this data will take much time, money and manpower.

Therefore, we offer the new method of estimations of environmental conditions, which can be applied in any part of Absheron peninsula – the use of space technologies in environmental diagnostics. We carried out environmental testing in appointed areas of Absheron peninsula therewith to correlate them with aerospace photography. Upon application of this method we shall obtain the reliable and efficient map of polluted territories.

8.1.1. Environmental Testing

Several types of contamination tests were conducted on soil and groundwater samples from potentially polluted sites in Absheron peninsula. Below you can see short description of them.

8.1.2. BTEX (Benzene, Toluene, Ethylbenzene and Xylene)

This group of volatile organic compounds (VOCs) is found in petroleum hydrocarbons and other common environmental contaminants. BTEX constitute serious threats to groundwater reservoirs and indoor climate deriving from contaminated sites due to the potential effects of benzene, which is considered a strong carcinogen, and which is highly mobile in the soil and groundwater environment.

8.1.3. PAH (Polycyclic Aromatic Hydrocarbons)

PAH present a group of over 100 different chemicals formed during the incomplete burning of coal, oil and gas, garbage, or other organic compounds. PAHs are known for their carcinogenic, mutagenic and teratogenic properties. PAHs can be found in petrochemicals, rubber, plastics, lubricants, antirust oil, paints, leather and other products.

8.1.4. TPH (Total Petroleum Hydrocarbons)

TPH is a term used to describe a broad family of several hundred chemical compounds that originally come from crude oil. The amount of TPH found in a sample is used as a general indicator of petroleum contamination at the site. TPH is released to the environment through accidents, as releases from industries, or as byproducts from commercial or private uses. The compounds in some TPH fractions can affect the blood, immune system, liver, spleen, kidneys, developing fetus, and lungs. Certain TPH compounds can be irritating to the skin and eyes.

8.1.5. Heavy Metals

Heavy metals that have been identified in the polluted environment include As, Cu, Cd, Pb, Cr, Ni, Hg and Zn. The presence of any metal may vary from site to site, depending upon the source of individual pollutant.

Soil contamination with heavy metals may cause changes in the composition of soil microbial community, adversely affecting soil characteristics.

8.1.6. Organochloride Pesticides

Organochloride pesticides are synthetic organic chemicals that have been used to control everything from fungus to grasshoppers. Polychlorinated biphenyls (PCBs) are by-products and constituents of a variety of industrial products, such as electrical transformers. Many organochloride pesticides are extremely persistent in the environment.

Because of chemical tests, we obtained results showing differentiation of contamination level in these points.

8.1.7. Advances of Use of Space Technologies

A Geographic Information System are in wide use as tools to digitize Remotely Sensed or cartographic data complemented with various ground - truth data which are geocoded using Global positioning systems. A Geographic Information System can help analyse the spatial characteristics of the data over various digital layers. If sequential data are available, quantification of spatial changes becomes possible through overlay analysis. A Geographic Information System is an expanding information technology for creating database with spatial information, which can be applied to both human settlements (e.g., demographic databases) and to the natural environment (e.g., distribution of populations and environmental factors). Most importantly, the combination of both types of database can ensure sustainable management. A Geographic Information System will continue to improve as an essential acquisition tool and analysis tool respectively not only in the analytical description of spatial subjects, but also in environmental planning, impact assessment, disaster management and simply monitoring Remote Sensing.

A Geographic Information System technology can be used for scientific investigations, resource management, asset management, archaeology, environmental impact assessment, urban planning, cartography, criminology, geographic history, marketing, logistics, prospectively mapping, and other purposes. The current challenge to Remote Sensing and

Geographic Information System – based on investigations is to combine data from the past and the present in order to predict the future.

Remote Sensing plays an integral role in environmental assessment. Remote Sensing will never replace the fieldwork and observations but it offers a great support in huge areas as follows:

- Remote and difficult access areas like dense forests, glaciated areas, swamps, high elevation, etc.;
- Areas undergoing rapid changes;
- Countries with poor infrastructure and limited transportation;
- Areas of active natural hazards and disasters: flooded areas, active volcanic regions, forest fires, earthquake and landslide hazardous areas, etc.; and
- Construction of a broad overview or a detailed map of a large area.
- Remote Sensing techniques can increase the speed in which one can analyse a landscape and therefore help make quick and focused decisions.

Among the available Remote Sensing technologies producing high spatial resolution data, aerial photography was superior to space - borne data, despite the higher spectral resolution of the latter. However, digital air - borne multi - spectral imagery such as the Compact Air - Borne Spectrographic Imager (CASI) is at least as accurate as aerial photography for the same purpose and it is less expensive to obtain and therefore more cost effective. It is also important to proceed in the evaluation of a new scientific application of more common imaging techniques such as video and photography from low - flying aircrafts. In space - borne Remote Sensing, the IKONOS satellite was the first one to challenge the very high spatial resolution (1m resolution) data obtained from air - borne Remote Sensing technology. The EROS satellite has a spatial resolution of 1.8 m but no multi - spectral capability. However, its future sensors are reported to generate multi - spectral combined with a spatial resolution of 0.82 m. In the meantime, imagery, the QUICKBIRD satellite leads the quality list of optical Remote Sensing with panchromatic imagery of 0.70 m spatial resolution and multi - spectral imagery of 3 m spatial resolution [332-334].

Achievements in information systems, satellites imaging systems and improved software technologies have led to opportunities for a new level of information products from Remote Sensed data. The integration of these new products into existing response systems can provide a wide range of analysis tools and information products on the base of developed Geographic Information System. Using the higher resolution of space imagery and change detection analysis, natural disaster awareness and damage assessment can be conducted rapidly and accurately [335-337].

8.1.8. Outcomes

Soil samples from six sites at Absheron Peninsula were collected and tested in environmental laboratory. Figure 45 below indicates those sites:

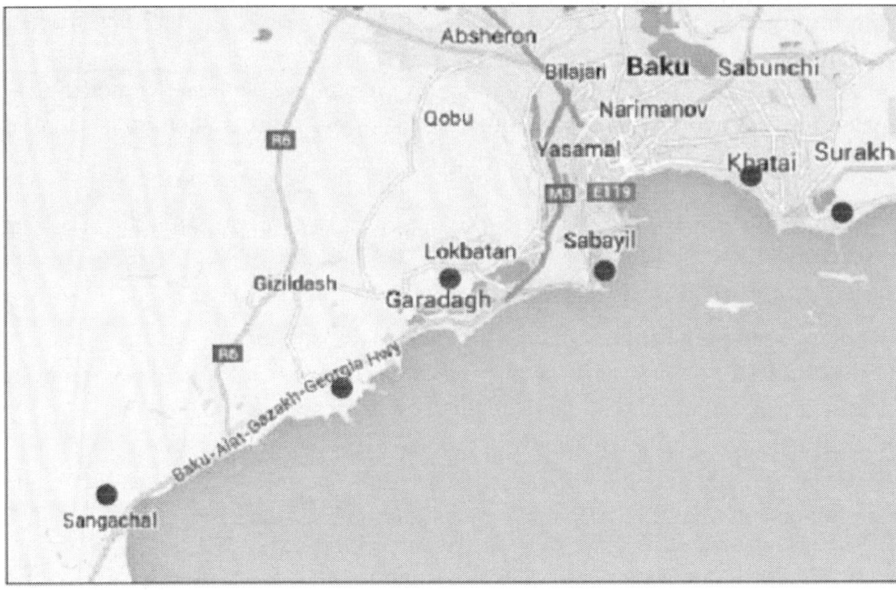

Figure 45. Sites where soils samples were collected.

Chemical analysis of soil contamination has been provided by different methods according to international standards. Table 22 demonstrates the results of conducted chemical tests, which have been integrated into the space data for further data processing and analysis.

Table 22. Contamination levels at different sites on Absheron peninsula

Contamination	Site No					
Level	1	2	3	4	5	6
None	*	*				
Low			*			
Medium						*
High				*	*	
Very High						

The laboratory test results have been based on the samples taken from different selected areas, which were coordinated using Global Positioning System. Indeed samples have been graded by status indicated above. The laboratory results were of merged with processed space data and applied and distributed for all selected territory in accordance with the oily soil polluted degree [338, 339].

This approach has been dedicated to the monitoring of environmental aspects of the soil condition. Two types of research of sources both laboratory chemical testing and use of Remote Sensing methods and Geographic Information System technologies for oil contamination degrees in the soil have been demonstrated. The successful use of space science and technology advances is reflected in oil pollution monitoring opening excellent advantages for rapidly soil monitoring investigation. It has been shown easy integration of laboratory testing data into the space with field investigation data as the instrument of problem solving in oil pollution exploration areas.

8.2. Advances of Space Technology in Geotechnical Studies

8.2.1. Abstract

Geographical Information System is the technology where geospatial data can be represented in the graphic form integrated into the geotechnical, geologic and hydrologic information routinely used by geotechnical engineers. A Geographical Information System makes available wide forms of spatial data to be integrated, selected and sorted with any number of physical, chemical or any possible environmental factors.

8.2.2. Introduction

There is no doubt that Remote Sensing method is the instrument for development of Geographical Information System. It is obvious that Geographical Information System technology developments need collection of initial data based on aero and space information. A Remotely Sensed information required for Geographical Information System technology applications needed to be processed depending on the kind of problem necessary to be solved.

Major areas of space science and technology applications in study of soil are following:

- Geotechnical Investigation;
- Foundations;
- Earthquake;
- Landslide Studies;
- Application of remote sensing and Geographical Information System;
- Geological and Geophysical Investigation;
- Industrial waste & contaminated soil.

Some of the important infrastructures in geotechnical investigations can be classified as [1]:

- Computer controlled Triaxial and consolidation Testing System;
- Computer controlled Laser Particle Size Analyser;
- Laboratory and field testing facilities for soil and fly ash deposit;
- Foundation Pile Diagnostic System;
- Geotechnical instrumentation for performance evaluation;
- Sub-surface profiling by Engineering Seismograph, Resistivity meter and Georadar;
- Strong Motion instruments;
- Digital Image Processing and Geographical Information System;
- Mobile drilling rig for soil and rock exploration;
- Petrological microscope;
- Total station for surveying & monitoring.

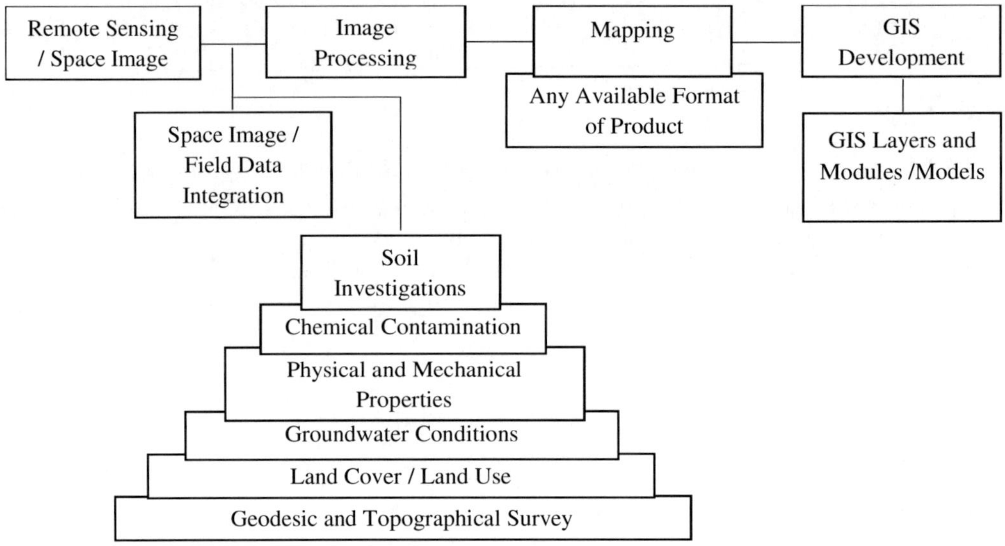

Figure 46. Geographical Information System developments in geotechnical studies.

8.2.3. Data Access in Geographical Information System Developments

To start from the segments of data collections based on Remote Sensing method use is an important stage of Geographical Information System developments (Figure 46). A quality of final product in geotechnical investigations depends on how correct and right Remote Sensing data shall be used for Geographical Information System technology application.

The fact is that the most expensive and time consuming component of Geographical Information System has been data access due to the lack of sufficient information in the area for the reason of limited store of geotechnical information in electronic format. There is a number of problems even with available information, which needed to be edited, as some objects on older maps have to be corrected and specified. Some paper maps can be scanned electronically as raster images, which convert map lines to a series of points and digits. Many Geographical Information System formulated to emphasize spatial relationships between mapped objects and a line usually represents such boundaries. The line may be a road, mapped boundary, or some sort of link between two other points of interest. Civil infrastructure elements, such as roads, may not be reflected accurately, in terms of absolute scale, but simply represented by a default line width(s) coded into the mapping software. This condition creates limitation in application mainly in Geographical Information System developments especially in dynamic change studies where necessary correct base information for correlation and comparison old and existing current situations.

8.2.4. Stage of Data Integration

A Geographical Information System makes possible to mix or integrate information that would otherwise be difficult to associate through other means. These could be soil chemical contamination, physical-mechanical properties in the form of scanned, geo-referenced and sandwiched with other kinds of data, such as topographic and geologic maps.

8.2.5. Data Structures

Digital geospatial data is collected and stored in many different formats. A Geographical Information System must be used to convert data from one type of structure to another. Satellite data can usually be "read" into the Geographical Information System in a raster format. Raster data files consist of rows of uniform cells coded according to data values. Computer can manipulate raster files quickly, but they are often less detailed and may be less visually appealing than vector data files. Vector digital data files have been captured as points, lines (a series of point coordinates) or areas (shapes bounded by lines). A typical vector file would be tax assessor's parcel maps [340, 341].

8.2.6. Data Modelling

A Geographical Information System allows two and three-dimensional characteristics of the Earth's surface, subsurface or atmosphere from geospatial data. Some common examples of data modelling would be creating isohyets based on different initial information sources. These data models can then be combined with other types of information layers in the Geographical Information System [342, 343]. Some common examples would be combining measured different information sources with elevation, or the thickness of a certain geologic formation (isopach) as compared to the depth to its upper surface (isopleth) where available.

8.2.7. Outcomes

One of the important issues of Geographical Information System is its ability to produce pleasing graphics that convey analyses to decision makers and the public at-large. These analyses usually begin with entering any codified restrictions, such as structural setbacks. An included attributes can then be electronically combined and weighted according to arbitrary values set by the body ordering the analysis. Such kind of developed hybrid data in the form of maps frustrate many engineers because they can arbitrarily be weighted to restrict or even eliminate development from areas where the project's detractors reside on adjacent parcels with all the same attributes.

8.2.8. Geotechnical Field Measurements

Field measurements in geotechnical investigations are very important in the development of Geographical Information System technology. A number of implementations is considered in the selected areas needed to be investigated [344, 345].

8.2.9. Drilling Works for Soil Samples

It has been selected for study the Sangachal district of Baku, the capital of Azerbaijan named after Shah Deniz. Drilling at the Shah Deniz flare area started immediately after realization of detailed geodetic and topographic investigation.

Drilling was conducted using auger with diameter 132mm, to the depth of 25 m (Drilling of BH-6 stopped at 18,0m depth). During the drilling process the whole information about the site, depth of boreholes, date and number of samples were recorded.

Figure 47. Laboratory geotechnical tests processing.

8.2.10. Sampling

Sampling was carried out during the borehole drilling using core sampler with internal diameter 100 mm. All possible measures to prevent cores from any kind of disturbance were undertaken. Samples were tightly packed with a suitable material into container. Each sample was marked with labels containing information about the site, borehole number, depth of the top and bottom of the drill run, date of sampling. Total amount of samples is 92.

Each sample sent to laboratory had marked and labelled duplicate which should be retained for period of 28 days after submission of approved Final Report.

8.2.11. Standard Penetration Test

This method covers the determination of the resistance to soils at the base of a borehole to the penetration of the split-barrel sampler when driven dynamically in a standard manner, and the obtaining of a disturbed sample for identification purposes.

The basis of the test consists of dropping with a free fall a hammer of mass 63,5kg on to a drive head from a height of 760mm. The number on blows (N) necessary to achieve a penetration of the split-barrel sampler of 300mm (after its penetration under gravity and below the seating drive) is regarded as the penetration resistance (N)

SPT were carried out in accordance with BS 5930 and BS 1377-9:1990. When conducting SPT 63,5kg hammer was used, height of fall - 75cm, dimensions of soil sampler: outside diameter 51mm, internal diameter 35mm.

The sampler assembly was lowered to the bottom of the borehole on the drive rods with the drive assembly on the top.

The initial penetration under total dead weight of assembly shall be recorded.

When the initial penetration exceeds 450 mm the seating drive and test drive shall be omitted and the "N" value recorded as 'zero.'

After the initial penetration, the test shall be carried out in two stages:

Seating drive: Using standard blows the seating drive shall be a penetration of 150 mm or 25 blows whichever is reached first.

Test drive: The number of blows required for further penetration of 300 mm. This is called the penetration resistance (N). If the 300 mm penetration cannot be achieved in 50 blows, the test drive shall be terminated.

The number of blows required to effect each 75 mm of penetration for both the seating and test drives shall be recorded. If the seating or test drive is terminated before the full penetration, the depth of penetration for the corresponding 25 blows and/or 50 blows respectively shall be recorded.

8.2.12. Soil Electrical Resistivity Test

Soil resistivity is dependent on moisture content and temperature as well as on soil constituents, so that it can vary seasonally and progressively due to the hydrological trends such as changing water tables or continuous drainage.

The accuracy of measurement is subject to a number of features, which should be borne in mind when assessing the implications of the value obtained.

Soil resistivity is generally measured by driving three equally spaced test spikes to a depth of up to 1m, the depth not exceeding 5% of their separation **a**. It is important to ensure that their resistance areas do not overlap. Current is passed between electrode X, the one being tested, and an auxiliary current electrode Y. The voltage drop between electrode X and a second auxiliary electrode Z is measured and the resistance of the electrode X is then the voltage between X and Z divided by the current flowing between X and Y. The source of current and the means of metering either the current and voltage or their ratio are often, but not necessarily, combined in one device.

8.2.13. Laboratory Processing of Field Data

There is no doubt that collected field data for geotechnical investigation require to be processed in laboratory condition. The following tests have been conducted for soil samples collected from investigated area.

8.2.14. Moisture Content

The water content is determined by drying selected moist/wet soil material (the mass of moist soil material is not less than 30g) for at least 18 hours to a constant mass in a drying oven at 105°C up to 110°C. The difference in mass before and after drying is used as the mass of the water in the test material. The mass of material remaining after drying is used as the mass of the solid particles. The ratio of the mass of water to the measured mass of solid particles is the water content of the material.

8.2.15. Particle Size Analysis

Particle size analysis can be performed by means of sieving and/or hydrometer readings. Sieving is carried out for particles that would be retained on a 0.063 mm sieve, while additional hydrometer readings may be carried out when a significant fraction of the material passes a 0.063 mm sieve.

In a sieve analysis, the mass of soil retained on each sieve is determined, and expressed as a percentage of the total mass of the sample. Prior to sieving, samples are treated with a dispersing agent (sodium hexameta phosphate), rinsed on a 0.063 mm sieve and dried.

The hydrometer method allows the measurement of the density of a suspension consisting of fine-grained soil particles and distilled water, to which a dispersion agent is added. This suspension is mixed using a high-speed stirrer. Testing is performed in a thermostatically controlled water bath (25° ± 0.5°).

The particle size is calculated according to Stokes' Law for a single sphere, on the basis that particles of a particular diameter were at the surface of the suspension at the beginning of sedimentation and had settled to the level at which the hydrometer is measuring the density of the suspension. These calculations require a value for the density of solid particles. Generally, a value of 2.65 t/m^3 is assumed. When other values are used, that is included in the laboratory report.

The hydrometer results for selected particles sizes are presented as a percentage of the total mass of the soil sample.

Particle size is presented on a logarithmic scale.

8.2.16. Maximum Dry Density and Optimum water Content (Proctor Test)

An indication of the state of compaction of a cohesion less (free-draining) soil is obtained by relating its dry density to its maximum and minimum possible densities (the limiting densities). The tests described in this section enable these parameters to be determined for cohesion less soils.

Two tests are described for determination of minimum density. One is a simple dry shaking test for sands and the other is a dropping test for gravelly soils.

8.2.17. Determination of Maximum Density of Sands

This test covers the determination of the maximum density to which a clean sand can be compacted. Method is suitable for sands containing a small amount of material passing a 63 mm test sieve, and up to 10% of fine gravel passing a 6.3 mm test sieve. The soil is compacted into a 1 L mould under water using an electric vibrating hammer.

8.2.18. Determination of Maximum Density of Gravelly Soils

This test covers the determination of the maximum density to which a gravel or sandy gravel can be compacted. The method is suitable for soils passing the 37.5 mm test sieve and containing a small amount of material passing the 63 mm test sieve. The soil is compacted into a 152 mm diameter CBR mould using an electrical vibrating hammer.

8.2.19. Derivation of Density Index

The density index, I_D, is an index which relates the void ratio, or dry density, of a soil sample, or the sample in situ, to the limiting voids ratios, or dry densities.

If the dry density of the soil in question is denoted by ρ_d, the density index I_D, may be calculated from the equation

$$\left(\frac{\rho d - \rho d\min}{\rho d\max - \rho d\min}\right)\left(\frac{\rho d\max}{\rho d}\right)$$

where

ρdmin is the dry density at the least dense state;

ρdmax is the dry density at the densest state.

8.2.20. Bulk Density

The bulk density of a soil, ρ is the mass per unit volume of the soil deposit including any water it contains. The dry density ρ_d is the mass of dry soil contained in a unit volume. Both are expressed in Mg/cm^3.

The method of immersion in water covers the determination of the bulk density and dry density of a sample of natural or compacted soil by measuring its mass in air and apparent mass when suspended in water. The method can be employed whenever lumps of material of suitable size can be obtained.

$$Vs = (mw - mg) - \left(\frac{mw - mf}{\rho d}\right)$$

where

V_s is the volume of the test specimen;

$\boldsymbol{m}_w$ is the mass of specimen and wax coating (in g);
$\boldsymbol{m}_g$ is the apparent mass of specimen and was coating when suspending in water (in g);
$\boldsymbol{m}_f$ is the mass of specimen after making up surface voids with filler (in g);
$\boldsymbol{\rho}_d$ is the density of paraffin wax (in g/cm^3).

$$\rho d = \frac{ms}{Vs}$$

where

$\boldsymbol{m}_s$ is the mass of the soil specimen.

$$\rho d = \frac{100\rho}{100 + \omega}$$

where

$\boldsymbol{\rho}_d$ is the dry density of the specimen (in Mg/m^3)
ω is the moisture content of the soil.
The bulk density and dry density are to express 0.01 Mg/m^3.

8.2.21. Atterberg Limits

Atterberg limits are determined on soil specimens with a particle size of less than 0.425 mm. If necessary, coarser material is removed by dry sieving. The Atterberg limits refer to arbitrarily defined boundaries between the liquid and plastic states (Liquid Limit, w_L), and between the plastic and brittle states (Plastic Limit, w_p) of fine-grained soils. They are expressed as water content, in percent.

The liquid limit is defined as the water content at which, a part of soil placed in a standard cup. As the next stage it was cut by a groove of standard dimensions which flow together with the base of the groove when the cup is subjected to 25 standard shocks. The four-point liquid limit test is usually carried out. Distilled water may be added during soil mixing to achieve the required consistency.

The plastic limit is defined as the water content at which, a soil can no longer be deformed by rolling into 3 mm diameter threads without crumbling.

The range of water contents over which the soil behaves plastically is the plasticity Index, I_p. This is the difference between the liquid limit and the plastic limit ($w_L - w_p$).

8.2.22. Unconsolidated–undrained Triaxial Compression Test

This test method covers determination of the strength and stress-strain relationship of a cylindrical specimen of either undisturbed or remoulded cohesive soil. Specimens are subjected to a confining fluid pressure in a triaxial chamber. No drainage of the specimen is permitted during the test. The specimen is sheared in compression without drainage at a constant rate of axial deformation.

8.2.23. Direct Shear Test

The soil is dried and sieved with 5 mm sieve, wetted and placed in the ring of the shear device. Three rings are prepared from one reformed sample, which is placed in the device for pre-consolidation. The applied loads varied in range 100kPa to 600kPa, then the box is filled with water or by natural moisture left for consolidation.

After consolidation the soil sample is quickly moved to the working box where applied load corresponds to the consolidation pressure. After 30 minutes of maintain of normal pressure (P), the sample is horizontally loaded (τ) in increments.

The value of one increment is equal to 5% of normal pressure. The horizontal pressure is applied with deformation speed of 0.05-0.08 mm/min.

The test is considered as completed if there is an instantaneous shear of the sample or total deformation of shear is 20% of sample diameter after application of scheduled horizontal pressure.

Key parameters that can be obtained from this test are angle of internal friction φ (grad) and cohesion C (kPa) determined from the plot τ = f (P) for three points. The shearing stresses are calculated by formula:

$$\tau = \frac{P}{A}$$

Where:

P- horizontal load;

A- sample area.

8.2.24. One-dimensional Consolidation Properties of Soils (Oedometer Test)

The oedometer test covers determination of the rate and magnitude of consolidation of a laterally restrained soil specimen, which is axially loaded in increments of constant stress until the excess pore water pressures have dissipated for each increment. Normally, each load increment is maintained for 24 hours.

The test is generally carried out on undisturbed cohesive specimens using a consolidometer (oedometer) apparatus, which is placed in a thermostatically controlled room (20°C). Selection of mounting method depends on soil characteristics. Soils, such as peat or over consolidated clays, are mounted dry. Moist sponges are placed in the oedometer cell to retain sample moisture conditions. Other samples are usually mounted using the wet mounting method. Disuntiled water is added to the cell when loads are applied to the loading arm. Key parameter that can be obtained from this test is the preconsolidation pressure ρ_c. The preconsolidation pressure is estimated using the graphical Casagrande construction. Other parameters that may be derived from this test are the compression index C_c.

8.2.25. Collapse Potential of Soils

This test method Covers the determination of the magnitude of one-dimensional collapse that occurs when unsaturated soils are inundated with fluid.

This test method is used to determine the magnitude of potential collapse that may occur for a given vertical (axial) stress and an index for rating the potential for collapse.

Collapse index I_e, percent – relative magnitude of collapse determined at 200 kPa.

Collapse potential I_c, percent – relative magnitude of soil collapse determined at any stress level other than 200 kPa.

The test method consists of placing a soil specimen at natural water content in a consolidometer, applying a predetermined applied vertical stress to the specimen with fluid to induce the potential collapse in the soil specimen. The fluid should be disentitled water when evaluating the collapse index, I_e. The fluid may simulate pore water of the specimen or other field condition as necessary when evaluating collapse potential I_c. The collapse potential versus severity of the problem is listed below in the Table 24:

Table 23. The list of potential versus severity collapse

Degree of collapse	Collapse Index I_e, %
None	0
Slight	0.1 to 2.0
Moderate	2.1 to 6.0
Moderately severe	6.1 to 10.0
Severe	>10

8.2.26. Determination of Permeability of Soils

The permeability of a soil is a measure of its capacity to allow the flow of water through the pore space between solid particles. The degree of permeability is determined by applying a hydraulic pressure gradient in a sample of saturated soil and measuring the consequent rate of flow. The coefficient of permeability is expressed as a velocity.

Test method includes determining the permeability of either undisturbed or disturbed and compacted to the given density samples. Disturbed sandy specimen must be dried before test.

Permeability of the sandy soil is determined with the constant hydraulic pressure gradient with water flow from bottom to the top of the specimen or from the top to the bottom. The specimen must be saturated from the bottom to the top before the application of hydraulic pressure. Permeability of the clayey soil is determined with the given pressure and variable hydraulic pressure gradient with water flow from the top to the bottom or from the bottom to the top. The specimen must be saturated from the bottom to the top without swelling prior to the test. This test procedure covers the determination of the coefficient of permeability using a permeameter. The volume of water passing through the soil in a known time is measured, and the hydraulic gradient is measured using manometer tubes.

This procedure is suitable for soils having maximum particle size 1/5 of inner diameter of permeameter.

The coefficient of permeability of about 1 cm/s is the approximate boundary between laminar and turbulent flow. The coefficient of permeability of about $1x10^{-4}$ cm/s is the approximate dividing between good drainage and poorly drained soils. The classification of soils according to permeability is as follows:

High degree of permeability:	k is over 0.1cm/s
Medium degree of permeability:	k is between 0.1 and 0.001
Low permeability:	k is between 0.001 and $1x10^{-5}$cm/s
Very low permeability:	k is between $1x10^{-5}$ and $1x10^{-7}$cm/s
Practically impermeable:	k is less than $1x10^{-7}$cm/s

8.2.27. Soil Chemical Analysis

Chemical analyses of soil is carried out to determine chloride, sulphate, calcium carbonate content, pH.

Soil pH is one of the most common measurements in soil laboratories. It reflects whether a soil is acid, neutral, basic or alkaline.

Soil pH may be arbitrarily described as follows: strongly acid (pH < 5.0), moderately to slightly acid (5.0 - 6.5), neutral (6.5 - 7.5), moderately alkaline (7.5 - 8.5), and strongly alkaline (> 8.5).

Depending on the amount of sulphate in contact with the concrete, it may be necessary to protect the concrete with a plastic liner, sulphate resistant concrete mix, or a protective adhesive coating.

8.2.28. Area Selected for Geotechnical Investigations

It has been selected the area of Sangachal district, Baku, the capital of Azerbaijan located on the coast of the Caspian Sea 45 kilometres (28 mi) south of Baku, Azerbaijan. There is the Sangachal Terminal, which is an industrial complex consisting of a natural gas processing plant and oil production plan. It makes very attractive and vital to monitor the area with contamination as more as possible information in point of environmental and ecological condition assessment. In this case, geotechnical investigations can play a significant role in collection and processing of data based on the use of modern technology applications.

Remote Sensing method and Geographical Information System technology is a key instrument in consideration of approaches of data collection for decision makers as well as suitable way in data storage for the future access.

Shah Deniz flare area is located adjacent to the Sangachal Terminal and lies within the Garadagh District, which includes Baku and then extends south along the Caspian coast to the south of Alyat. The Garadagh District was established in 1923 and comprises five city settlements including Lokbatan which is the District's administrative center. The four communities in the immediate vicinity of the Terminal (Sangachal Town, Umid, Masiv 3 and Azim Kend) are likely to be the most directly affected by the socio-economic impacts of Shah Deniz flare project.

8.2.29. Data Processing Stage

Figure 48 shows the location of area selected for geotechnical investigations. It reflects of points (red colour) conducted measurements into the mapping system. The number of points required to be investigated on geotechnical parameters definitions depend on the engineering task and scope of work reflected in the project requirements. This circumstance finds the accuracy of conducted engineering service. In some cases, it becomes very expensive in conducting huge numbers of geotechnical investigations. It relates to the scale of area needed to be investigated and type of engineering facilities intended to be constructed.

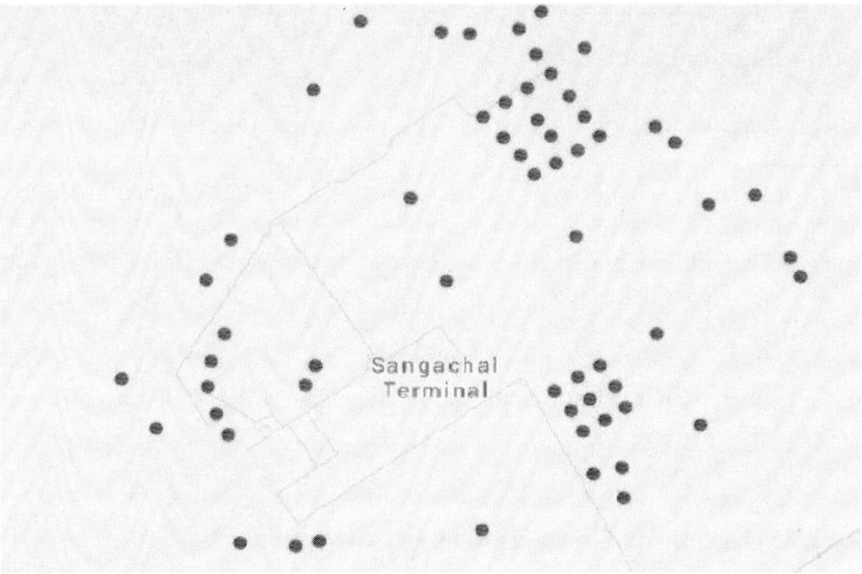

Figure 48. Map of the selected area.

Sensitive area has been demonstrated in the map from the point of view of engineering importance of conducted geotechnical measurements. As seen from the Figure there are a big number of ground samples have been taken from the area.

The definition of the process of Geographical Information System development in geotechnical investigations is highly important. Figure 48 describes segments of data collection from different sources and those integration [346].

It has been considered to develop layers for Geographical Information System which contains the following performance:

- Soil investigation data;
- Chemical contamination;
- Physical and mechanical properties of soil;
- Ground water condition of soil;
- Land Use/Land Cover;
- Geodetic and topographic data.

A more geotechnical data can be integrated into Geographical Information System layers depending on the requirements of engineering solutions for the selected area. The main available information is tied to various forms of georeferenced information. The accurate merging of topographic map with space image for selected area is required for achieving demanded cartographic corrections. An excellent advantage of Geographical Information System is its incorporate processing subroutines that can transform older data to modern coordinates if a sufficient number of georeferencing points can be co-located on both the old and new maps. These georeference points may be established and constructed benchmarks (state or installed), old structures, roads, or even aboveground power lines; anything that can be identified on both maps in the Geographical Information System.

The results of data imposed on the space image for the selected area is shown in Figure 49. The space image with spatial resolution of 1m has been used in the figure. It is enough high resolution, which is more than enough for the case of geotechnical studies of the selected area.

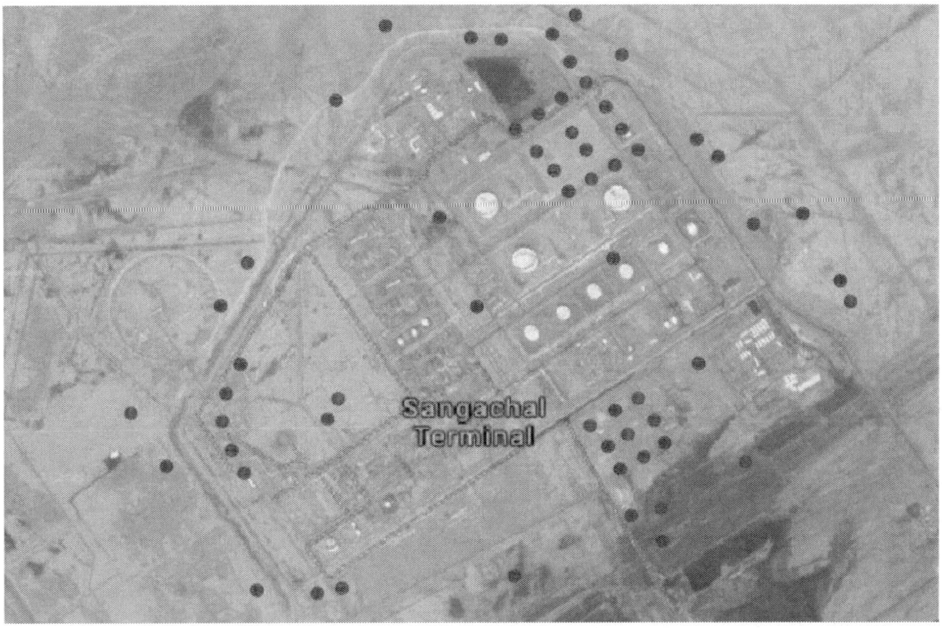

Figure 49. Integration of geotechnical data and processed space image data.

8.2.30. Geotechnical Data As an Indicator for Environmental Studies

The impact on the land from an environmental point can play a significant role in their use as an indicator. Geotechnical methods application in study of environment is one of the possibilities for monitoring of the processes especially in modification of indicators making available correct identification of the reason of consequences of environmental impacts. The best way to apply this indicator is evaluating the benefits for the land, stabilization, improvement of the quality of the soil, creation of a better area to sustain biodiversity (Flora and Fauna), works for their protection, and improving the sustainable character for the future uses. Subsequently, the influence of the wide possible negative consequences can be analysed from an economical point of view.

There is no doubt that all indicators in environmental investigation needed to be accurately identified in the beginning stage of measurements. It is necessary to undertake the behaviour of the selected area for investigation such as feature of existing flora and fauna as well as soil condition. It makes possible to select the segments of geotechnical samples and measurements. It gives opportunity to manage successfully all the stages of investigation processes. It will contribute time spent and economic aspects of expected implementations based on the geotechnical methods.

There is one more advantage to use Remote Sensing methods and Geographical Information System technology applications integrating all data collected by different sources of data. At the same time, it makes possible to archive and store accessed data in e-version by layouts with reflection of data segments, which is the most convenient in use during any needs. Remote Sensing and Geographical Information System technology application open wide advances in study of dynamic changes of environment and design required cases and identification of reasons both positives and negatives in processing stages of data.

An integration of geotechnical and space technology advances is an excellent instrument. As far the field data is highly important in space technology data processing as one of the main and necessary stages of the whole process. Geotechnical data is an excellent evidence in field measurement access integrated further into the space data.

This approach describes the geotechnical investigations of the selected area in Absheron peninsula, Azerbaijan. It contains the measurements of geotechnical data such as:

- standard penetration test;
- soil electrical resistivity test;
- laboratory processing of filed data;
- moisture content;
- particle size analysis;
- maximum dry density and optimum water content (proctor test);
- determination of maximum density of sands;
- determination of maximum density of gravelly soils;
- derivation of density index;
- bulk density;
- atterberg limits;
- unconsolidated–undrained triaxial compression test;
- direct shear test;
- one-dimensional consolidation properties of soils (oedometer test);
- collapse potential of soils; and
- Determination of permeability of soils.

There is no doubt that it is an excellent source reflecting soil condition which can be used for engineering solution in any stage of implementation, tender package preparation, design and construction.

At the same time, the integration of geotechnical data into the Geographical Information System developed on the base of space data collected by method of Remote Sensing is an advantage of application of outcomes in a wide areas of engineering such as project coordination and management, construction, supervision where the use of suitable and simplicity of data access is required.

In the meantime, the use of the data integrated into Geographical Information System makes possible to link the existing information to the coordinate system, which is very important in all stages of achievements.

Undoubtedly, the performance of the integration of geotechnical data into the space information accessed by means of Remote Sensing with further development of geographic information system, which demonstrates the variety of use and success in the area. It is not limited only with measurements of the selected area by use of geotechnical methods in collection of data for the purpose of further engineering activities. At the same time, it is approach of different problems solution, for instance, environmental as well as integration of data into the space information.

8.3. Remote Sensing and Geographic Information System in the Power Energy Transmission – Environmental Problems Solution

There is no doubt that the selection methodology and appropriate technology is required for any problem solving. It relates for environmental issues in power energy transmission.

Space-related innovation includes adapting existing Earth-systems modelling data into different resources assessments and technological diffusion of tools such as Geographic Information Systems and Global Positioning System devices. These tools can help collection and interpretation of necessary data to develop more integrative models [347, 348].

Remote Sensing method as a key base of further applications can also provide reliable database for decision makers. There is luckily integrations of specialist in the space technology and authorities for clarification of situation and decisions. The Space technology Remote Sensing can also enable measurement and monitoring land use and land cover changes for classifications.

How the spectral signature analysis provides expected results on power energy transmission to the environmental issues? It can be used an excellent indicator for power energy transmission impact studies. Healthy vegetation contains large quantities of chlorophyll, which gives more vegetation to its distinctive green colour. In referring to healthy crops, reflectance in the blue and red parts of the spectrum is low since chlorophyll absorbs this energy; in contrast, reflectance in the near-infrared (NIR) spectral regions is high. Stressed or damaged crops experience a decrease in chlorophyll content and changes to the internal leaf structure. The reduction in chlorophyll content results a decrease in reflectance of the green region and internal leaf damage results a decrease in near-infrared reflectance. These reductions in green and infrared reflectance provide early detection of plant stress.

The detection of chlorophyll content and greenness in plant is the basis behind some vegetation indices, including the Normalized Difference Vegetation Index (NDVI), the Wetness component of the Tasseled Cap Transformation and the Normalized Difference Index. We can consider each of these indices depending on the cases required for application.

One of the important issues is the definition of the element or elements of space technology applications in power energy transmissions. It is required to find out necessary and sensitive indicators needed to be monitored using Remote Sensing based on space images. There are no other sources for direct fixing and registration of the power energy impact on the environment. It demands the selection of some indirect indicators, for instance animals, population or vegetation. As far as the vegetation can be assumed as a high sensitive indicator to the power energy transmission. It would be created a positive approach of use of vegetation for monitoring of environment investigation purposes.

For carrying out of the goals following approaches have been undertaken:

- The use of ALOS space imagery to create the land use / land cover basic map for the investigated area using urban, agriculture, garden, scrub, open area, river, river crossings, stream, canal, road, railroad basic classes;
- The use of Landsat ETM space imagery for detection of potentially impacted areas by power line transmissions selected for investigated district of Azerbaijan using a tasseled cap transformation;
- To evaluate the sensitivity of each approach to be characterized by the power energy transmissions through the statistical tests involving comparison of influenced areas extracted from an inventory of soils and a geomorphology maps.

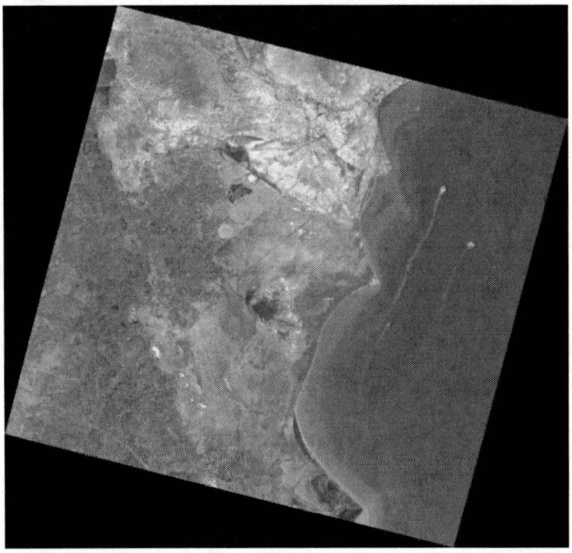

Figure 50. ALOS satellite image.

ALOS imagery was acquired on June 10, 2007 with the spatial resolution 10 m. Figure 1 demonstrates used ALOS imagery for investigations. At the same time, one Landsat Enhanced Thematic Mapper (ETM) satellite image for analysis was selected. The image was georeferenced to UTM zone 39 North, WGS84 using a first degree polynomial rectification algorithm with 25 ground control points (GCPs) extracted from a digitized topographic map at the scale of 1:100 000. The image pixels were resampled to 28.5 x 28.5 m using a nearest-neighbour interpolation method to preserve radiometric integrity. The root mean square (RMS) error obtained in the rectification process was less than 1 pixel (28.5 m).

8.3.1. Area Selection for Study

It has been taken attention to the selection of investigated area, which was highly sensitive issue for the success of the study. It was undertaken circumstances of low influences of human and other impacts than power energy transmission to the environment of the selected area. Figure 2 reflects most suitable area chosen for study. It is obvious that there are eliminations related to the access as well as seasons and years of space images. It is especially sensible for the countries with the low of nature of use space technology for different purposes.

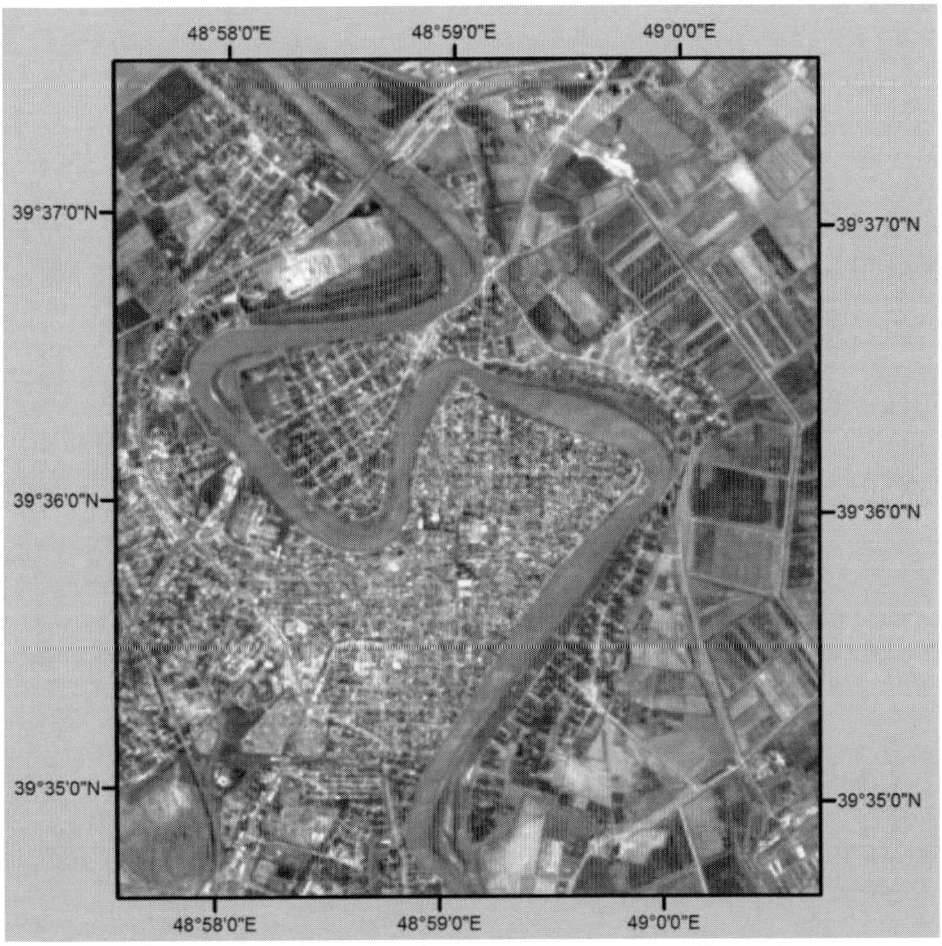

Figure 51. ALOS imagery of the selected area.

It is necessary to mention that selected area consists of a number of features for study which are going to be considered the available changes affecting by power energy transmission. Those elements of legend can be successfully used as an indicators for monitoring of environmental issues.

8.3.2. Space Image Processing

For the reason of identification of power energy transmission impact to the environment, it was necessary to select indicators for monitoring. It is indirect classification that making available to study environmental aspects of the investigations for the selected area. There is no doubt that in this case vegetation or fauna and flora can be a key element for those studies. Taking into account the suitability of this approach the space image was classified between following general classes (Figure 52):

1. Urban or Built-up Land
2. Agricultural Land
3. Garden
4. Scrub
5. Open area
6. River
7. Stream
8. Canal
9. Road
10. Railroad

8.3.3. Tasseled Cap Transformation

A tasseled cap transformation was applied to the image processing to be able to optimize data viewing [349, 350]. The tasseled cap transformation offers three types of data structures axes that can be used to define vegetation information content:

Brightness: a weighted sum of all bands defined in the direction of the principal; variation in soil reflectance.

Greenness: orthogonal to brightness, a contrast between the near-infrared and visible bands that strongly related to the amount of green vegetation in the scene.

Wetness: relates to canopy and soil moisture and effective to discriminate wet areas [351].

Obviously, there is a demand to compare existing statistical data and current processed data for dynamic change studies of the selected area impacted by the power energy transmissions. At the same time, it is necessary to carry out permanent monitoring and collect data (statistical and space data) with further processing in order to identify clearly existing condition of the selected area for study of power energy transmissions impact on environment.

The tasseled cap algorithm using coefficients for ETM imagery are:

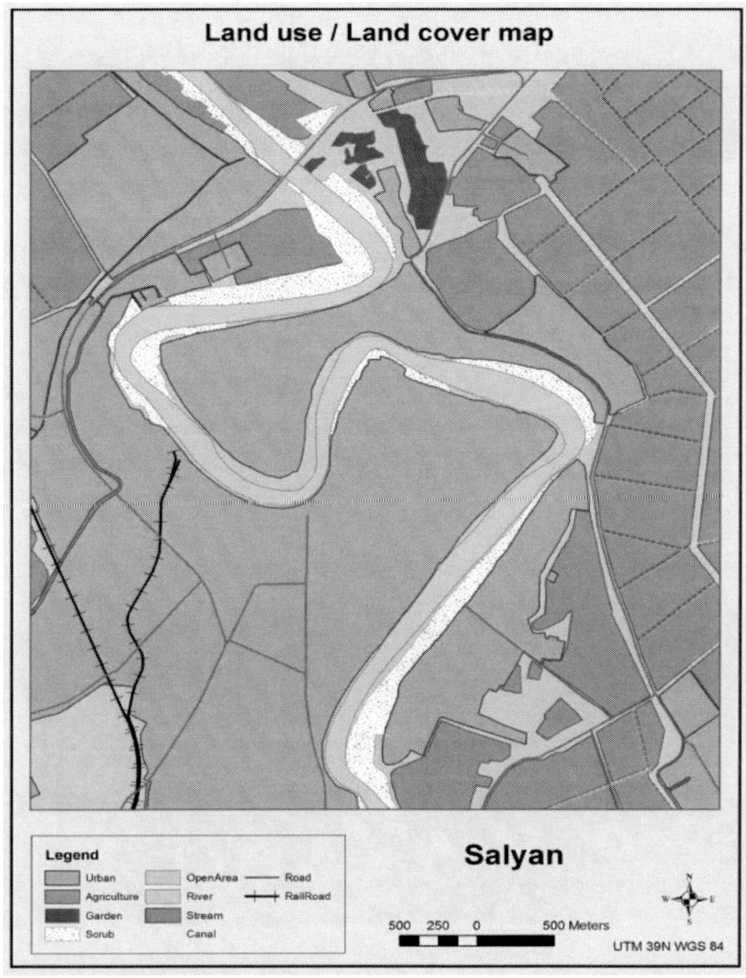

Figure 52. Land use/Land cover map.

First Landsat-7 image was converted to at-satellite radiance using equation (1),

$$L_{sat} = \frac{L_{\max sat} - L_{\min sat}}{DN_{\max} - DN_{\min}} \cdot \left(DN - DN_{\min}\right) + L_{\min sat}$$

where,

$L_{\max sat}$ – is band-specific spectral radiance scaled to $DN_{\max}$ (W/ (m2*ster*μm),

$L_{\min sat}$ – is band-specific spectral radiance scaled to $DN_{\min}$ (W/ (m2*ster*μm),

$DN_{\max}$ – is maximum quantized calibrated digital number (255), and

$DN_{\min}$ – is minimum-quantized calibrated digital number (0 for LPGS data, 1 for NLAPS data).

Equation (1) accounts for gain state (i.e., high/low setting) by using the respective published LMIN/LMAX values (Landsat 7 Science Data User's Handbook).

After conversion to satellite radiance, the image was exposed to Tasseled Cap transformation. The Tasseled Cap operation (also known as Kauth's Tasseled Cap) computes the three Kauth biophysical indices (greenness, brightness, and wetness) from raster objects that contain the six spectral bands of ETM imagery: ETM1, ETM2, ETM3, ETM4, ETM5, ETM7. This spectral information is translated into values that represent site's biophysical properties. The process produces three output raster objects that represent greenness, brightness, and wetness using a set of linear combinations of Landsat ETM spectral bands (Figure 53).

a) RGB Composite with bands 5, 4, 3

b) RGB Composite with layers brightness, greenness and wetness

Figure 53. (Continued).

c) Pseudo colour composite with wetness index

Figure 53. a) The original ETM image; b) The image applied the tasseled cap transformation; c) The derived wetness layer.

It opens an opportunity to compare data passed and current stage of situation in point of view of environmental aspects of power energy transmission impact. It is needed to consider an issue that numbers of collected old and current data directly reflect the accuracy and reliability of the applied methodology. There are always eliminations of access to the old data statistical and other type of information often for the reason of luck of such a data. It is felt especially in the stage of the space technology applications in such impacted areas as power energy transmissions.

An application advance of modern space technology for study of any aspects including power energy transmissions environmental influences is the useful instrument for the future undertakings. In this paper, the methodology of approach of space technology on study of power energy influence in the environment is demonstrated. It is just beginning stage of attempt for such investigations needed to be raised in the future explorations.

8.4. Human Impact of Environmental Problem Solving by Advances of Space Technology

Oil spill detection is sunlit performed mainly by visual observation, which is limited to favourable sea and atmospheric conditions and any operation in rain, fog, or darkness is eliminated. Visual observations are restricted to the registration of the spill because there is no mechanism for positive oil detection. Very thin oil sheens are also difficult to detect especially in misty or other conditions that limit vision. Oil is difficult to discover in high seas and among debris or weeds where it can blend in to dark backgrounds such as water, soil or shorelines. A huge number of naturally occurring substances or phenomena can be mistaken for spilled oil. These include sun glint, wind shadows and wind sheens, biogenic or natural oils from fish and plants, glacial flour (finely, ground mineral material usually from glaciers) and oceanic or revering fronts where two different bodies of water meet. The usefulness of visual observations is limited, however, it is an economical way to document spills and provide baseline data on the extent and movement of the spilled oil [352].

Estimation of the quantity of oil observed at sea is the main issue for the detection of the oil spill. Observers are generally able to distinguish between sheen and thicker patches of oil. However gauging the oil thickness and coverage is not always easy and it can be more difficult if the sea is rough. It is essential to view all such estimates with considerable caution.

8.5. What We Can Do for More Achievements and Challenges in the Oil Spill Detection?

Purpose of the Remote Sensing equipment mounted in aircraft is increasingly used to monitor, detect and identify sources of illegal marine discharges and to monitor accidental oil spills. Remote Sensing devices except infrared video and photography from airborne platforms, thermal infrared imaging, airborne laser fluouro sensors, airborne and satellite optical sensors use satellite Synthetic Aperture Radar (SAR). Advantages of SAR sensors over optical ones are their ability to provide data in poor weather conditions and during darkness. Remote Sensing method operates detecting properties of the surface such as colour, reflectance, temperature or roughness of the area. Spilled oil can be detected on the surface when it modifies one or more of these properties. Cameras relying on visible light are widely used and may be supplemented by airborne sensors which detect oil outside the visible spectrum and are thus able to provide additional information about the oil. The most commonly applied combinations of sensors include Side - Looking Airborne Radar (SLAR) and downward - looking thermal infrared and ultraviolet detectors or imaging systems.

8.6. Advances of Remote Sensing in Oil Spill Detection

Environmental pollution, including oil spill is one of the major ecological problems. Negative human impacts demand the development of appropriate legislations within the national and international framework for marine and coastal environment as well as the onshore protection. Several seas, for instance the Mediterranean, the Baltic and the North Seas were declared as special areas where ship discharges are completely prohibited [353]. The Caspian Sea as a closed ecosystem is also considered as danger polluted the water basin for fauna and flora.

In this regard, the environmental protection of the Caspian Sea has a priority status for Azerbaijan as a closed water basin ecosystem. This area, as a highly sensitive area in the World requires permanent ecological monitoring services where oil and gas from the subsurface of the Caspian Sea has been developing almost more than a century. This status of the Caspian Sea is expected to be retention at least for the coming fifty years.

Usefulness: Remote Sensing is a useful method in several modes of oil spill control, including a large-scale area of surveillance ability, specific site monitoring and advantages of technical and technological assistance in emergency cases. There is a significant capacity of providing essential information to enhance strategic and tactical decision - making, decreasing response costs by facilitating rapid oil recovery and ultimately minimizing impacts.

Observation can be undertaken visually or by using Remote Sensing systems. In Remote Sensing, a sensor other than human vision or conventional photography is used to detect or map oil spills.

Advantages: Remote Sensing plays an integral role in environmental assessment. Remote Sensing will never replace the field work and observations but it offers a great support in a huge areas as follows:

- Remote and difficult access areas like dense forests, glaciated areas, swamps, high elevation, etc.;
- Areas undergoing rapid changes;
- Countries with poor infrastructure and limited transportation;
- Areas of active natural hazards and disasters: flooded areas, active volcanic regions, forest fires, earthquake and landslide hazardous areas, etc.; and
- Construction of a broad overview or a detailed map of a large area.

It is obvious that for any application of Remote Sensing methods the identification of spatial resolution of the space images is necessary. It depends on the task required to be considered for study and development. Spatial resolution requirements are various but it is necessary to consider even for massive oil spills. It is well known that spills at sea from windrows with widths are often less than 10 m. A spatial resolution is greater than it is required to detect these spills. Furthermore, when considering oil spills, information is often required on a relatively short timescale to be useful to spill response personnel.

Figure 54. ENVISAT ASAR image in the Caspian Sea near the Absheron peninsula for oil spill due to the offshore oil production.

Figure 54 shows an example of oil spill of the Absheron peninsula taken by ENVISAT ASAR. This figure reflects a necessity of the permanent monitoring of the Caspian Sea for more sensitive areas.

8.7. Geographic Information System Developed on the Base of Remote Sensing Data

A Geographic Information System are in a wide use as tools to digitize Remotely Sensed or cartographic data complemented with various ground - truth data which are geocoded using Global Positioning Systems. A Geographic Information System can help analyse the spatial characteristics of the data over various digital layers. If sequential data are available, quantification of spatial changes becomes possible through overlay analysis. A Geographic Information System is an expanding information technology for creating database with spatial information which can be applied to both human settlements (e.g., demographic databases) and to the natural environment (e.g., distribution of populations and environmental factors). Most importantly, the combination of both types of database can ensure sustainable management. A Geographic Information System will continue to improve as an essential acquisition tool and analysis tool respectively not only in the analytical description of spatial subjects, but also in environmental planning, impact assessment, disaster management and simply monitoring Remote Sensing [354].

The integration of space imagery with Geographic Information Systems allow accurate geo - positioning of pipeline vector information to the local land use and topography representation becomes a very useful planning and decision support tool. Location of the linear elements infrastructure can be placed as a vector file over a one - meter spatial resolution satellite image and coloured red. Sensitive environmental areas are then identified as green through a land classification analysis on the Geographic Information System product. Location of other elements on the surface like roads, agriculture areas and infrastructures are also clearly distinguishable in case of the availability of the high spatial resolution of space imagery [355].

The main disadvantage of some new Remote Sensing technologies for instance, as IKONOS is their commercial nature and the very high process charged are a limiting factor that prevents the scientific community to access these data. This is particularly true for developing countries and for the republics of the former Soviet Union, the government of which may bear the high cost of aerial surveys. But make the photographs available at a marginal cost to the local academic institutions, far below the price of satellite imagery with the same resolution.

Many publications exist on the effects of global change for marine environments, including oil spill effects. Apart from the integration of past and present data scientists should put a lot of efforts to the prediction of future scenarios and the establishment of early warning systems in order to help guarantee the survival of sustainable ecosystems.

The study and investigation of oil spill is important and became suitable and useful instrument in study of ecosystem. Application of the space science and technology advances for oil spill impact is making available comprehensive approach of an exploration of oil spill impact. It is possible to inquire not only detection of oil spill, measure of sizes of impacted area and forecasting of further consequences of damages but also is to study and consider of oil spill impact on flora and fauna of sea. There are especially benefits and advantages in the close ecosystem like the Caspian Sea.

CONCLUSION

The environmental aspects of engineering activities and development have been studied in this Chapter. There are methods and methodology describing environmental impacts in different cases of engineering executions. At the same time, Remote Sensing method and Geographic Information System applications in environmental problem solving are demonstrated.

REFERENCES

[1] Remote Sensing and Geographic Information System Systems RS-GIS, http://www.set.ait.ac.th/page.php?fol=rsgis&page=rsgis.

[2] An Introduction to Remote Sensing & GIS Introduction, Yale University Genocide Studies Project, Remote Sensing & GIS Research.

[3] Gleason, Art, Scott Kaiser, and Tamara Smith 1994 Center for Earth Observation Users Guide. 8th Revision August 2004 by Larry Bonneau, Yale University.

[4] Lillesand, Thomas M., Ralph W. Kiefer, and Jonathan W. Chipman 2004 Remote Sensing and Image Interpretation, Fifth edition. Wiley, New York.

[5] Sabins, Floyd F. 1997 Remote Sensing: Principles and Interpretation, Third edition. W. H. Freeman and Company, New York.

[6] Schowengerdt, Robert A. 1997 Remote Sensing: Models and Methods for Image Processing. Academic Press, New York.

[7] The Association for Geographical Studies Remote Sensing and Geographical Information System (GIS) Dr. Punyatoya Patra Associate Professor, Aditi Mahavidyalaya, University of Delhi pp.1-28.

[8] The Application of Remote Sensing and GIS in Urban Planningmsc. Eng. Bahaa Eddin I. Al Haddad.

[9] Wiliam L. Stefanov, Michael S. Ramsy, Philip R. Christensen "Monitoring Urban Land Cover Change: An Expert System Approach to Land Cover Classification of Semiarid to Urban Centres Department of Geological Sciences, Centre for Environmental Study and Department of Geology and Planetary, Arizona State University and University of Pittsburgh, USA.

[10] P.S.Uttarwar, Join Director (Planning) Application of GIS and Remote Sensing in Urban Planning, Implementation and Monitoring of Urban Projects – case study of Rohini and Dwarka project, New Delhi, Delhi Development Authority, New Delhi, India.

[11] Prof. Binyi Liu Toward The 21ST Century: GIS and Remote Sensing Application on Architecture, Urban and Landscape Panning. College of Architecture and Urban Planning, Tongji University, China.

[12] Jerry C. Coiner Enterprise GIS in Urban Planning and Management Ph.D., Principal Remote Sensing and Information Systems Co., Kailua-Kona, HI USA, http://www.rsisco.com/Enter_GIS.htm.

[13] Study On Land Use Suitability Assessment Of Urban-Rural Planning Based On Remote Sensing—A Case Study Of Liangping In Chongqing Luo Lingjun, He Zong,

Hu yan (Chongqing Geomatics Center, The Fourth Floor of No.D, Neptune Building, Xinguang Avenue, High-tech Devolopment Zone, Chongqing New North Zone).

[14] Jiao Li-min, Liu Yao-lin, Application of Fuzzy Neural Networks to Land Suitability Evaluation. *Geomatics and Information Science of Wuhan University.* Vol.29-NO.6, 2004.

[15] Zhou Zhong-fa. The rural land suitability assessment of karst areas based on GIS. *Journal of Guizhou Normal University* (Natural Sciences), Vol 24-NO.5, 2006.

[16] Lv Yun-feng etc, On Land Suitability Evaluation Based on Remote Sensing and GIS. *Journal of Changchun Teachers College,* Vol.26-NO.5, Oct, 2007.

[17] Liu Wu etc, Land Adaptability Evaluation Based on Vector-Raster Mixed Data Model. *Resources and Environment in the Yangtze Basin,* VOl.15-NO.3, May, 2006]

[18] Yang Shu-jia etc, The Method Research of Basic Farmland Distribution Based on The Land Suitability Assessment. *Journal of Shandong Normal University* (Natural Science), Vol.22-NO.3, March, 2007]

[19] Tang Xian-ming etc, A Study on the Site Selection Model of Mountain Town - Take Wushan County as Example. *Journal of Mountain Science,* Vol.19-NO.2, April 2001.

[20] Ling Yun-chuan, Study on Land suitability evaluation theory and method. *Modern agricultural science and technology,* Vol.18, 2007.

[21] Shi Tong-guang etc, Progress in Research on Land Suitability Evaluation in China. *Progress in Geography,* Vol.26-NO.2, March 2007.

[22] Hoffer, R., (1978). "Biological and Physical Considerations in Applying Computer-aided analysis techniques to Remote Sensor data," Remote Sensing: The quantitative approach, edited by P. H. Swam and S. M Davis U. S. A., McGraw-Hill.

[23] Adia S., Rabiu A., (2003). *Change detection of vegetation cover, using multi-temporal remote sensing data and GIS techniques, Geospatial* Application Papers, Environment.

[24] Remote Sensing and Geographical Information System (GIS) Panyatoya Patra.

[25] Qiming Zhou, "The Integration of GIS and Remote Sensing for Land Resource and Environment Management" *Proceedings of United Nation ESCAP Workshop on Remote Sensing and GIS for Land and Marine Resources and Environment Management, School of Geography,* The University of New South Wales Sydney, NSW 2052, Australia. February 13-17, 1995, pp.43-55.

[26] G.D. Squires, *"Urban Sprawl and the Uneven Development of Metropolitan America" Urban Responses.* Urban Institute Press, Washington, D.C.2002, pp.1- 22.

[27] Fei Yuan, Kali E Sawaya, Brian C. Loeffelholz, Marvin E. Bauer "Land cover classification and Change analysis of the Twin Cities (Minnesota) Metropolitan Area" *By multitemporal Landsat remote Sensing, Remote Sensing of Environment,* Vol. 99, August 2005, pp. 317-327.

[28] Claes Andersson, Steen Rasmussen and Roger White, "Urban settlement Transitions" *Environment and Planning,* Vol.29, 2002, pp.841-865.

[29] E.F Lambin, M.D.A. Rounsevell, and H.J.Geist, "Are agricultural land-use Models able to predict changes in land use intensity" *Agriculture, Ecosystems and Environment,* Vol. 82, 2000, PP.321– 331.].

[30] Jensen, J.R. (1996). "Introductory digital processing": *A remote sensing perspective* (2n edition). Upper Saddle River, NJ: Prentice-Hall.

[31] Prof. S. Ramachandran Application of Remote Sensing and GIS.

[32] Bastin. J. (1988). Measuring areas of coral reefs using satellite imagery, Symposium on Remote Sensing of the Coastal Zone, *Gold Coast Queensland.* P vii.1.1-vii.1.9.

[33] Ramachandran, S., Devasenapathy, J., Sundramoorthy, S. and Krishnamoorthy, R. (2000a). Satellite Remote Sensing Application in Coastal Zone Management, In: *Marine Remote Sensing Applications.Institute for Ocean Management,* Anna University, pp. 87-90.

[34] Nayak. S. R, Chauhan, P., Chauhan, H.B., Balamurugan, A. and Nath, A.N. (1996). IRS-1C Applications for Coastal Zone Management. *Current Science,* 70 (7): 614-618.

[35] Hussin. Y.A, Mahfud, M. and Zuhair Michael Weir (1999). Monitoring Mangrove Forests using Remote Sensing and GIS. GIS development proceedings, ACRS.

[36] IGBP (1994). *Land-Ocean Interactions in the Coastal Zone*, Report No. 33. p74. SAC, (1992). Coastal Environment. Space Applications Centre (SAC) Scientific Note (RSAM/SAC/COM/ SN/11/92), Ahmedabad, India, 114 pp.

[37] Krishnamoorthy. R, and Ramachandran, S. (2000). Application of Remote Sensing in Mangrove Wetland Studies. In: *Marine Remote Sensing Applications.* Institute for Ocean Management, Anna University, pp. 168-205.

[38] Krishnamoorthy. R, Bhattacharya, A. and Natarajan, T. (1993). Mangrove and Coral Reef Mapping of South Andaman Islands through Remote Sensing, In: *Sustainable Management of Coastal Ecosystems.* M.S.

[39] IOM report (2003b). *Integrated Coastal Zone Management Plans for Andaman and Nicobar Islands.* Submitted to Ministry of Environment and Forests, Govt. of India.

[40] IOM report (2001). No Impact Zone Studies in Pulicat Lake. Submitted to Dept. of Ocean Development, Govt. of India.

[41] IOM report (2003a). *Identification of Ecologically Important Areas in India.* Submitted to Ministry of Environment and Forests, Govt. of India.

[42] Krishnamoorthy. R. (1995). Remote Sensing of Mangrove forest in Tamil Nadu Coast, India. Ph.D thesis, Anna University, Madras, 202 pp.

[43] Ramachandran, S., Ramesh, S. and Krishnamoorthy, R. (2000b). Application of Remote Sensing and GIS in Coastal Lagoonal Ecosystem: A Case Study from Pulicat Lake, Southern India, In: *Marine Remote Sensing Applications.* Institute for Ocean Management, Anna University, pp. 333-343.

[44] Lillesand. T.M. and Kiefer, R.W. (1987). Remote Sensing and Image Interpretation. John Wiley & Sons. Mumby. P.J, Green, E.P., Clark, C.D. and Edward, A.J. (1998). Digital Analysis of Multispectral Airborne Imagery of Coral Reefs. *Journal of the International Society for Reef Studies,* 17: 59-69.

[45] Ramachandran. S. (1993). Coastal Zone Information System – Pilot project for Rameswaram area. Report submitted to Department of Ocean Development. Govt. of India, 40 pp.

[46] Ramachandran. S, Krishnamoorthy, R., Sundramoorthy, S., Parviz, Z.F., Kalyanamuthiah, A. and Dharanirajan, K. (1997). Management of Coastal Environments in Tamilnadu and Andaman & Nicobar Islands based on Remote Sensing and GIS approach. MAEER'S MIT, *Pune Journal,* IV (15 & 16), Special issue on Coastal Environmental Management, pp. 129-140.

[47] Ramachandran. S, Sundramoorthy, S., Krishnamoorthy, R., Devasenapathy, J. and Thanikachalam, M. (1998). Application of Remote Sensing and GIS to Coastal

Wetland Ecology of Tamilnadu and Andaman and Nicobar group of Islands with special reference to Mangroves. *Current Science,* 75(3): 101-109.

[48] R. Kancheva, D. Borisova1, G. Georgiev, Yu. Tishchenko, Test-Sites in Remote Sensing Studies and Earth Observations, *Journal Fundamental Space Research,* 2009.

[49] Main Report Socio-Economic Benefits Analysis of GMES, GMES Benefits and Impacts Report, Prepared by Price Water House Coopers, ESA Contract Number 18868/05, October 2006.

[50] *Official Journal of the European Union,* 22.2.2007, L 54/30-80, Corrigendum to Council Decision 2006/971/EC of 19 December 2006 concerning the specific programme 'Cooperation' implementing the Seventh Framework Programme of the European Community for research, technological development and demonstration activities (2007 to 2013), (Official Journal of the European Union L 400 of 30 December 2006).

[51] R. Kancheva, H. Nikolov, D. Borisova, 2005, Modeling and verification in vegetation spectral studies. *Annual of UMG "St. Ivan Rilski,"* Part I: Geology and Geophysics, vol. 48, Sofia, Publishing House "St. Ivan Rilski," 221-224.

[52] R. Kancheva, et al., 2005, Modeling and verification in vegetation spectral studies. *Annual of UMG "St. Ivan Rilski,"* Part I: Geology and Geophysics, vol. 48, Sofia, Publishing House "St. Ivan Rilski," 221-224.

[53] http://www.azerbaijan.orexca.com/azerbaijan_regions.shtml.

[54] M Raghunath, *Application of Remote Sensing and GIS in Urban Land Suitability Modeling at Parcel Level Using Multi-Criteria Decision Analysis,* Indian Institute of Remote Sensing National Remote Sensing Agency Dept. of Space, Govt. of India Debra Dun-248 001, Bangalore, India, 2006.

[55] Abeid, J., Allouche, E., Arditi, D., and Hayman, M. (2003). "Photo-Net II: A Computer based monitoring system applied to project management." *Autom. Constr.,* 12(5), 603-616.

[56] Adjei-Kumi, T., Retik, A., and Shapira, A. (1996). "Integrating on-site tasks into planning and scheduling of construction projects." Procccdings of the International Symp. On Organization and Management of Construction, Shapping Theory and Practice, Managing the Construction Project and Managing Risk: D.A. Langford and A.retik, eds., E & FN Spon, Glasgow, U.K., 283-292.

[57] Anbazhagan.S, Dept.of Earth Sciences, IIT Bombay, "GIS and Its Applications" edited in 2003.

[58] CSRE, IIT Bombay, "Geographic Information System principles and applications." *One day workshop,* March 2005.

[59] Cheng, M.Y., and Chen, J.C. (2000). "Integrating barcode and GIS for monitoring construction projects." *Autom. Constr.,* 11(1), 23-33.

[60] Cheng, M.Y., and Yang, S.C. (2001). "GIS- based cost estimates integrating with material layout planning." *J. Constr. Eng. Manage.* 127(4), 291-299.

[61] Cheng, M.Y., and O'Connor, J.T. (1996). "ArcSite: Enhanced GIS for Construction site layout." *J. Constr. Eng. Manage.,* 122(4), 329-336.

[62] Dr.F.Escobar, Assoc. Prof. G. Hunter, Assoc. Prof. I.Bishop, Dr.A.Zeger, *"Introduction to GIS,"* http://www.sli.unimelb.edu.au/gisweb.

[63] Dr. G.S. Rao, Ajita Kini, Saumyagjit Roy, Anwar Hussan, *"IT and GIS – a complementary technology blend"* ESRI 2005. "ArcGIS Solutions for Civil Engineering GIS—the Intelligent Platform for Civil IT."

[64] Kamat, V.R., and Martinez, J.C. (2002). *"Comparison of Simulation-driven construction operations visualization and 4D CAD."* Exploring New Frontiers: Proc:., 2002 Winter Simulation Conf., E. Yucesan, C.H. Chen, J.L. Snowdon, and J.M. Charnes, eds., Institute of Electricity and Electronics Engineers, New York, 1765-1770.

[65] Kang-tsung Chang, "Introduction to Geographic Information Systems," 2006, Tata McGraw-Hill KLG Systel, *Course Material on Primavera Enterprise Project Management* (P6), 3-8

[66] Koo, B., and Fischer, M. (200). "Feasibility study of 4D CAD in commercial construction." *J. Constr. Eng. Manage.,* 126(4), 251-260.

[67] P.Venkatachalam, B.Krishna Mohan, GRAM++ -Indigenous GIS Software Suite, CSRE, IIT Bombay. http://www.csre.iitb.ac.in/gram++/Publications.htm.

[68] P. Venkatachalam, B.Krishna Mohan, M. Farooque and Amit Kotwal, GRAM++ Viewer - A User Friendly Tool for Spatial Planning, CSRE, IIT Bombay. http://www.csre.iitb.ac.in/gram++/Publications.htm.

[69] P.Venkatachalam, Principles of GIS, gram++ GIS Package Development and Applications, CSRE, IIT Bombay http://www.csre.iitb.ac.in/gram++/Publications.htm.

[70] P. Venkatachalam, Principles and Applications of GIS and Image Processing, Lecture Notes, One day Workshop organized by CSRE, IIT Mumbai, April 2007.

[71] P.S. Gahlot, B.M. Dhir, Construction Planning and Management PMBOK, Project Management - Body of Knowledge as defined by the Project Management Institute — PMI.

[72] Retik, A., and Shapira, A. (1999). "VR based planning of Construction projects." *IEEE Comput. Graphics Appl.*, 24(1), 6-12.

[73] Geoff Zeiss, Director of Technology, Autodesk, Inc., "Convergence: Integrating Geospatial Technology and Architectural and Engineering Design."

[74] Geoff Zeiss, Director of Technology, Autodesk, Inc., *"Worldwide Challenges facing the Construction Industry."*

[75] Harold Kerzner, Ph.D, Project Management – A Systems Approach to Planning, Scheduling and Controlling Hemal Lad, Geographic Information System & Its application in site layout work, M. Tech Dissertation VJTI Mumbai, 2000.

[76] J.Patrick Moore, "Building a Baseball Stadium using GIS," *A Case study of IntegralGIS Company.*

[77] Retik, A. (1997). "Planning and monitoring of construction projects using virtual reality." *Proj. Manage J.,* 97(1), 28-31Research Futures: Academic responses to industry challenges. (1997), Construct IT Center of Excellence, Salford, U.K.

[78] Stephen E. Poku and Prof. David Arditi, M.ASCE, "Construction Scheduling and Progress Control Using GIS," *Journal of Computing in civil engineering,* Vol.20, No. 5, Sept.1 2006.

[79] Shankar Bagale (2006), *"Planning and Management of Irrigation projects using Remote Sensing and GIS Technology."* NRSA, 15-22.

[80] Spatial India ESRI India News, Noida, Friday, December 2, 2005, *Special supplement on Eight Annual India ESRI User Conference.* "From Pilots to enterprise Deployments." 1-2.

[81] Vijay K. Bansal, Mahesh Pal, "GIS Based Projects Information System for Construction Management," *Asian Journal of Civil Engineering* (Building and Housing) Vol.7, No.2 (2006) Pages 115-124.

[82] Remote Sensing & GIS For Planning Studies, Dr. Talat EKSIOGLU, P.O. Box 53179, Riyadh 11583, KSA, talateksioglu@ags-group.com.

[83] Dahdouh, F. G., 2002, *"The Use of Remote Sensing and GIS in the Sustainable Management of Tropical Coastal Ecosystems."*

[84] Montaya, L., 2002, "GIS and Remote Sensing in Urban Disaster Management." *5th AGILE Conference on Geographic Information Science,* Palma (Balearic Islands, Spain).

[85] Coleman, T.L. 1992, "Three-dimensional modeling of an image-based System Applications Product to aid land planning." *Geocarto* 4: 47-52.

[86] Garry, G., 1992, *Lusage des photographies aériennes, volume 3 of Serie Environnement et Aménagement* (Paris: Les Editions du STU).

[87] Ehlers, M., Jadkowski, M. A., Howard, R. R., And Brousten, D. E., 1990. "Application of SPOT data for regional growth analysis and local planning." *Photogrametric Engineering and Remote Sensing,* 56, pp. 175-180.

[88] Eiumnoh, A., 2000, "Integration of Geographic Information Systems (GIS) and Satellite Remote Sensing (SRS) For Watershed Management"

[89] http://rst.gsfc.nasa.gov/Sect1/Sect1_17.html.

[90] http://rst.gsfc.nasa.gov/Sect1/Sect1_16.html.

[91] http://rst.gsfc.nasa.gov

[92] *"Engineering and Engineering Management – Space Technology Application"* Nargiz A. Babayeva, Rustam B. Rustamov.

[93] Jeljeli, M. N., Russel, J. S., Meyer, H. W. G., and Vonderohe, A. P. ''Potential application of geographic information systems to construction industry.'' J. Constr. Engrg. and Mgmt., *ASCE,* 119(1), 72–86, (1993).

[94] Oloufa, A. A., Eltahan, A. A., and Papacostas, C. S. ''Integrated GIS for construction site investigation.'' J. *Constr. Engrg. and Mgmt.,* ASCE, 120(1), 211–222, (1994).

[95] Cheng, M. Y., and Yang, S. C. ''GIS-Based cost estimates integrating with material layout planning.'' *J. Constr. Engrg. and Mgmt.,* ASCE, 127(4), 291-299, (2001).

[96] Li, H., Kong, C. W., Pang, Y. C., Shi, W. Z., and Yu, L. ''Internet-based geographic information systems system for E-Commerce application in construction material procurement.'' J. Constr. Engrg. and Mgmt., *ASCE,* 129(6), 689–697, (2003).

[97] Zhong, D., Li, J., Zhu, H., and Song, L., ''Geographic information system based visual simulation methodology and its application in concrete dam construction processes.'' J. Constr. Engrg. and Mgmt., *ASCE,* 130(5), 742–750, (2004).

[98] Rustam B. Rustamov, Sefer Kurnaz, Saida E. Salahova, Sabina N. Hasanova, Maral H. Zeynalova, *Application of Space technology with Engineering Management for Pollution Monitoring*, 2011, Published by Nova Science Publishers, Inc. New York, USA, ISBN: 978-1-61324-238-4.

[99] Space Technology Applications – Cost Effectiveness Arif Hashimov, Sabina N. Hasanova, Sevinj R. Rustamova, Kamran T. Mammadov, Rustam B. Rustamov.

[100] Rustam B. Rustamov, Saida E. Salahova, Sabina N. Hasanova, Maral H. Zeynalova, "Space Technology as the Tool in Climate Change Monitoring System," Climate Change. *Research and Technology for Adaptation and Mitigation,* Edited by Juan

Blanco and Houshan Kheradmand, Publishers INTECH, ISBN 979-953-307-277-6, August, 2011.

[101] Sefer Kurnaz, Rustam B. Rustamov, Maral Zeynalova, Saida E. Salahova, Monitoring of the impacts of the natural disaster based on the use of space technology, *KSAS International Journal,* Vol. 10, No 1, May 2009.

[102] Sefer Kurnaz, Rustam B. Rustamov, Saida Salahova, Maral Zeynalova, River Inundation Impact Based on Space Technology Applications, Proceedings of 4th International Conference on Recent Advances in Space Technologies, *IEEE,* Istanbul, Turkey, June 11-13, 2009, pp. 265-272.

[103] Director, NNRMS-RRSSC, ISRO Headquarters.

[104] Chidambaram, M., "Scientific GIS Digital Base Maps - Urban Planning using GIS / RS Technologies.

[105] Michael Goodchild, David J, David W Rhind, Paul A, (1990) *"GeographicalInformation System, Management issues and Applications,"* John Wiley &Sons, Vol 2.

[106] Priyadarshi, N. (2004). Distribution of arsenic in Permian coals of NorthKaranpura coalfield, Jharkhand. Jr. Geol. Soc. India, vol.63, pp. 533-536.

[107] Sarkar, S. & Kanungo, D.P., "Landslides in Relation to Terrain Parameters- a Remote Sensing and GIS approach.

[108] Srirama, B., Bhatt, M.R. & Pathan, S.K., "Least Cost Highway AlignmentUsing GIS Technique."

[109] http://spie.org/x22909.xml.

[110] *International Journal of Civil Engineering,* Vol. 6, No. 4, December 2008 235 M. H. Sebt, E. Parvaresh Karan , M. R. Delavar.

[111] DeMers, M. N, (2002). *Fundamentals of Geographic Information Systems,* 3rd Ed., John Wiley and Sons, Inc.

[112] Jeljeli, M. N., Russel, J. S., Meyer, H. W. G., and Vonderohe, A. P. (1993). ''Potential application of geographic information systems to construction industry.'' *J. Constr. Engrg. and Mgmt.,* ASCE, 119(1), 72–86.

[113] Oloufa, A. A., Eltahan, A. A., and Papacostas, C. S. (1994). ''Integrated GIS for construction site investigation.'' *J. Constr. Engrg. and Mgmt., ASCE,* 120(1), 211–222.

[114] Cheng, M. Y., and Yang, S. C. (2001). ''GIS-Based cost estimates integrating with material layout planning.'' *J. Constr.Engrg. and Mgmt., ASCE,* 127(4), 291-299.

[115] Li, H., Kong, C. W., Pang, Y. C., Shi, W. Z., and Yu, L. (2003). ''Internet-based geographic information systems system for e-Commerce application in construction material procurement.'' *J. Constr. Engrg. and Mgmt., ASCE,* 129(6), 689–697.

[116] Zhong, D., Li, J., Zhu, H., and Song, L., (2004). ''Geographic information system based visual simulation methodology and its application in concrete dam construction processes.'' *J. Constr. Engrg. and Mgmt., ASCE,* 130(5), 742–750.

[117] Poku, S. E., and Arditi, D. (2006). ''Construction scheduling and progress control using geographic information systems.'' J. *Comput. Civil. Engrg., ASCE,* 20(5), 351–360.

[118] Cheng, M. Y., and O'Connor, J. T. (1996). ''ArcSite: Enhanced GIS for construction site layout.'' *J. Constr. Engrg. and Mgmt., ASCE,* 122(4), 329–336.

[119] Elbeltagi, E., Hegazy, T., and Eldosouky, A. (2004). ''Dynamic layout of construction temporary facilities considering safety.'' *J. Constr. Engrg. And Mgmt., ASCE*, 130(4), 534–541.

[120] Choi, B., and Flemming, U. (1996). "Adaptation of a layout design system to a new domain: Construction site layouts." *Proc. of 3rd Congress Computing in Civil Engineering,* ASCE, A/E/C Systems, Anaheim, Calif., 718–725.

[121] Foster, D., Hollingsbead, M., Johnson, B., Kinzy, S., Purcell, K., and Tudos, J. (1989). Life cycle solution for infrastructure management, McDonnell Douglas, St. Louis, Mo.

[122] Tommelein, I. D., Levitt, R. E., and Hayes-Roth, B. (1992a). ''Site layout modeling: How can artificial intelligence help?'' *J. Constr. Engrg. and Mgmt., ASCE,* 118(3), 594–611.

[123] Yeh, I.-C. (1995). ''Construction-site layout using annealed neural network.'' *J. Comput. Civ. Eng., ASCE,* 9(3), 201–208.

[124] Hegazy, T. M., and Elbeltagi, E. (1999). "EvoSite: Evolution-based model for site layout planning." *J. Comput. Civ. Eng., ASCE,* 13(3), 198–206.

[125] Al-Hussein, M., Niaz, M. A., Yu, H., and Kim, H. (2006). ''Integrating 3D visualization and simulation for tower crane operations on construction sites.'' *Automation in Construction.* 15, 554–562.

[126] Zhang, P., Harris, F. C., Olomolaiye, P. O., and Holt, G. D. (1999). ''Location optimization for a group of tower cranes.''*J. Constr. Engrg. and Mgmt., ASCE,* 125(2), 115–122.

[127] Zayed, T. M., and Minkarah, I. (2004). ''Resource allocation for concrete batch plant operation: Case study.'' *J. Constr. Engrg. and Mgmt., ASCE,* 130(4), 560–569.

[128] Sadeghpour, F., Moselhi, O., and Alkass, S. T. (2006). ''Computer-aided site layout planning.'' J. Constr. Engrg. and Mgmt., ASCE, 132(2), 143–151.

[129] Occupational Safety and Health Administration (OSHA). (2003). "Safety and health regulations for construction." 29 Code of federal regulation, Part 1926, Washington, D.C.

[130] El-Rayes, K., and Khalafallah, A. (2005). ''Trade-off between safety and cost in planning construction site layout.'' *J. Constr. Engrg. and Mgmt., ASCE,* 131(11), 1186–1195.

[131] Tommelein, I. D., Levitt, R. E., and Hayes-Roth, B. (1992b). "SightPlan model for site layout" *J. Constr. Eng. Manage., ASCE, 118*(4), 749–766.

[132] Tommelein, I. D., Levitt, R. E., Hayes-Roth, B., and Confery, T. (1991). "SightPlan experiments: Alternate strategies for site layout design." *J. Comput. Civ. Eng., ASCE,* 5(1), 42–63.

[133] Flemming, U., and Chien, S. F. (1995). "Schematic layout design in SEED environment." *J. Archit. Eng., ASCE,* 1(4), 162–169.

[134] Harmanani, H., Zouein, P., and Hajar, A. (2000). "An evolutionary algorithm for solving the geometrically constrained site layout problem." *Proc., of the 8th Int. Conf. on Computing in Civil Engineering, ASCE,* Reston, Va., 1442–1449.

[135] Leung, W. T. A., and Tam, C. M. (1999). ''Prediction of hoisting time for tower cranes for public housing construction in Hong Kong.'' *Constr. Manage. Econom.* 17(3), 305–314.

[136] Li, H., and Love, P. E. D. (1998). "Sitelevel facilities layout using genetic algorithms." *J. Comput. Civ. Eng., ASCE*, 12(4), 227–231.

[137] Mawdesley, M. J., Al-jibouri, S. H., and Yang, H. (2002). "Genetic algorithm for construction site layout in project planning." *J. Constr. Engrg. and Mgmt., ASCE,* 128(5), 418–426.

[138] Philip, M., Mahadevan, N., and Varghese, K. (1997). "Optimization of construction site layout: A genetic algorithm approach." *Proc., 4th Congress on Computing in Civil Engineering*, ASCE, Reston, Va., 710–717.

[139] Tam, C. M., and Tong, T. K. L. (2003). "GA-ANN model for optimizing the locations of tower crane and supply points for high-rise public housing construction." *Constr. Manage. Econom.,* 21(3), 257–266.

[140] Tam, C. M., Leung, W. T. A., and Liu, D. K. (2002). ''Nonlinear models for predicting hoisting times of tower cranes.'' *J. Comput. Civil. Engrg., ASCE,* 16(1), 76–81.

[141] Zouein, P. P., Harmanani, H., and Hajar, A. (2002): "Genetic algorithm for solving site layout problem with unequal-size and constrained facilities." *J. Comput. Civ. Eng., ASCE,* 16(2), 143–151.

[142] GEOENSEÑANZA. Vol.11-2006 (1). Enero - junio. p.17- 28. ISSN 1316-60-77 1V. K. Bansal and Mahesh Pal. Department of Civil Engineering. National Institute of Technology, Kurukshetra, 136119 Haryana, India. Email: mpce_pal@yahoo.co.uk, 18 V. K. Bansal and Mahesh Pal Geographic Information Systems for Construction Industry: A Methodology to Generate 3-D View of Buildings.

[143] CHANG, Kang-Tsung. (2002). *Introduction to Geographic Information Systems,* Tata McGraw-Hill, New Delhi.

[144] CHENG, M.Y., and YANG, C.Y. (2001). "GIS-Based cost estimate integrated with material layout planning." *Journal Construction Engineering and Management, ASCE,* 127 (4), 291-299.

[145] BANSAL, V. K., and PAL, M. (2005)." GIS in Construction Project Information System" Proc. Map India, 8th annual international conference and exhibition in the field of GIS, GPS, Arial Photography, and Remote Sensing. New Delhi, India.

[146] Densham P J, Rushton G 1988 Decision support system for location planning. In Golledge R G, Timmermans H (eds) Behavioural modelling in geography and planning. Beckenham, Croom Helm: 56–90.

[147] SUN, W., and HASELL, M.J. (2002). "Exploring a GIS prototype to improve the management of the architectural design, engineering and construction building product process." Environment Science and Research Institute (ESRI) Conference, http://www.gis.com/library/userconf/proc02/pap0212/p0212.

[148] CAMP, C. V., and BROWN, M. C. (1993). "GIS procedure for developing three-dimensional subsurface profile." *Journal of Computing in Civil Engineering, ASCE,* 7 (3), 296-309.

[149] Oloufa, A. R., Eltahan, A. A., and Papacostas, C. S. (1994). "Integrated GIS for construction site investigation." *Journal Construction Engineering and Management, ASCE,* 120 (1), 211-222.

[150] Cheng, M.Y., and O'connor, J.T. (1994). "Site layout of construction temporary facility using enhanced-Geographic Information System (GIS)." *Automation in Construction,* 3, 11-19.

[151] Cheng, M.Y., and Chen, J.C. (2002). "Integrating barcode and GIS for monitoring construction progress" *Automation in Construction,* 11, 23-33.

[152] LI, H., Chen, Z., Yong, L., and Kong, S.C.W. (2005). "Application of integrated GPS and GIS technology for reducing construction waste and improving construction efficiency" *Automation in Construction,* 14, 323-331.

[153] LI, H., Kong, C.W., Pang, Y.C., Shi, W.Z., and YU, L. (2003). "Internet-based geographic information Geographic Information Systems for construction Industry: A methodology to generate 3-d view of buildings GEOENSEÑANZA. Vol.11-2006 (1). Enero - junio. p.17 - 28. ISSN 1316-60-77.

[154] VARGHESE, K., and O'CONNOR, J.T. (1995). "Routing large vehicles on industrial construction site*." Journal Construction Engineering and Management, ASCE,* 121 (1), 1-12.

[155] Cheng, M.Y., and Chang, G.L. (2001). "Automating utility route design and planning through GIS" *Automation in Construction,* 10, 507-516.

[156] Zhong, D., Li, J., Zhu, H., and Song, L. (2004). "Geographic information system based visual simulation methodology and its application in concrete dam construction processes." *Journal Construction Engineering and Management, ASCE,* 130 (5), 742-750.

[157] V. K. Bansal and Mahesh Pal systems system for E-commerce application in construction material procurement" *Journal Construction Engineering and Management, ASCE,* 129 (6), 689-697.

[158] Lo, C.P., Yeung, A.K.W. (2002*). Concept and Techniques of Geographic Information System,* Prentice-Hall of India, New Delhi.

[159] ArcView GIS (1996), *Reference Manual for Window by Environment Science and Research Institute*. Redlands, C.A.

[160] Cheng, M.Y., and O'CONNOR, J.T. (1996). "ArcSite: Enhanced GIS for construction site layout." *Journal Construction Engineering and Management, ASCE,* 122 (4), 329-336.

[161] Clark, K. (2001). *Getting Started with Geographic Information Systems,* Upper Saddle River, Prentice Hall N.J.

[162] A G-O Yeh, Urban planning and GIS, pp. 877-888 http://www.geos.ed.ac.uk/~gisteac/gis_book_abridged/files/ch62.pdf.

[163] Sinton D F 1977 *"The user's guide to IMGRID, an information manipulation system for grid cell structures.* Cambridge (USA), Department of Landscape Architecture, Harvard University.

[164] Webster C J 1994 GIS and the scientific inputs to urban planning. Part II – prediction and prescription. *Environment and Planning B: Planning and Design* 21: 145–57.

[165] Bardon K S, Elliott C J, S*tothers N 1984 Computer applications in local authority planning departments 1984: a review.* Birmingham, Department of Planning and Landscape, City of Birmingham Polytechnic.

[166] Campbell H J 1994 How effective are GIS in practice? A case study of British local government. *International Journal of Geographical Information* Systems 8: 309–25.

[167] French S P, Wiggins L L 1990 California planning agency experiences with automated mapping and geographic information systems. *Environment and Planning B: Planning and Design* 17: 441–50.

[168] Yeh A G-O 1991 The development and applications of geographic information systems for urban and regional planning in the developing countries. *International Journal of Geographical Information Systems* 5: 5–27.

[169] Han S Y, Kim T J 1989 Can expert systems help with planning? *Journal of the American Planning Association* 55: 296–308.
[170] Berry J K 1987 Fundamental operations in computer-assisted map analysis. *International Journal of Geographical Information Systems* 1: 119–36.
[171] Tomlin C D 1990 Geographic information systems and cartographic modeling. Englewood Cliffs, Prentice-Hall Webster C J 1993 GIS and the scientific inputs to urban planning. Part I – description. *Environment and Planning B: Planning and Design 20*: 709–28.
[172] Hopkins L D 1977 Methods for generating land suitability maps: a comparative evaluation. *Journal of the American Planning Association* 44: 386–400.
[173] McHarg I 1969 Design with nature. New York, Doubleday Newton P W, Taylor M A P (eds) *1986 Microcomputers for local government planning and management.* Melbourne, Hargreen.
[174] Steinitz C, Parker P, Jordan L 1976 Hand-drawn overlays: their history and prospective uses. *Landscape Architecture* 66:444–55.
[175] Levine J, Landis J D 1989 Geographic information systems for local planning. *Journal of the American Planning Association* 55: 209–20.
[176] Marble D F, Amundson S E 1988 Microcomputer based geographic information systems and their role in urban and regional planning. *Environment and Planning B: Planning and Design* 15: 305–24.
[177] Royal Town Planning Institute (1992) *Geographic information systems: a planner's introductory guide prepared by the Institute's GIS Panel.* London, The Royal Town Planning Institute.
[178] Newton P W, Taylor M A P, Sharpe R (eds) 1988 Desktop planning: microcomputer applications for infrastructure and services planning and management. Melbourne, Hargreen Paulsson B 1992 Urban applications of satellite remote sensing and GIS analysis. Washington DC, The World Bank.
[179] Goodchild M F, Parks B O, Steyaert L T (eds) *1993 Environmental modeling with GIS.* New York, Oxford University Press.
[180] Barnsley M J, Barr S L, Hamid A, Muller J P A L, Sadler G J, Shepherd J W 1993 Analytical tools to monitor urban areas. In Mather P M (ed.) *Geographical information handling – research and applications.* Chichester, John Wiley & Sons: 147–84.
[181] Douven W, Grothe M, Nijkamp P, Scholten H 1993 Urban and regional planning models and GIS. In Masser I, Onsrud H (eds) *Diffusion and use of geographic information technologies.* Dordrecht, Kluwer: 317–37.
[182] Yeh A G-O, Li X 1996 Urban growth management in the Pearl River Delta: an integrated remote sensing and GIS approach. *ITC Journal:* 77–85.
[183] Longley P, Higgs G, Martin D 1994 The predictive use of GIS to model property valuations. *International Journal of Geographical Information Systems* 8: 217–35.
[184] Schüller J 1992 GIS applications in environmental planning and assessment. Computers, Environment and Urban Systems 16: 337–53.
[185] Shiffer M J 1995c Interactive multimedia planning support: moving from standalone systems to the World Wide Web. *Environment and Planning B: Planning and Design* 22: 649–64.

[186] Armstrong M P, Densham P J, Lolonis P, Rushton G 1992 Cartographic displays to support locational decision-making. *Cartography and Geographic Information Systems* 19: 154–64.

[187] Yeh A G-O, Chow M H 1996 An integrated GIS and location–allocation approach to public facilities planning. *Computers, Environment and Urban Systems* 20: 339–50.

[188] Chuvieco E 1993 Integration of linear programming and GIS for land-use modelling. *International Journal of Geographical Information Systems* 7: 71–83.

[189] Landis J D 1995 Imagining land use futures: applying the California urban futures model. *Journal of the American Planning Association* 61: 438–57.

[190] Despotakis V K, Giaoutzi M, Nijkamp P 1993 Dynamic GIS models for regional sustainable development. In Fischer M M, Nijkamp P (eds) *Geographic information systems, spatial modelling, and policy evaluation.* Berlin, Springer: 235–61Urban planning and GIS.

[191] Eastman J R, Kyem P A K, Toledano J, Jin W G 1993 GIS and decision making. Worcester (USA), *Clark Labs for Cartographic Technology and Geographic Analysis.*

[192] Campbell H J, Masser I 1995 GIS and organisations: how effective are GIS in practice? London, Taylor and Francis Carver S J 1991 Integrating multi-criteria evaluation with geographical information systems. *International Journal of Geographical Information Systems* 5: 321–39.

[193] Armstrong M P, Densham P J 1990 Database organization strategies for spatial decision support systems. *International Journal of Geographical Information Systems* 4: 3–20.

[194] Armstrong M P, Densham P J, Rushton G 1986 Architecture for a microcomputer-based decision support system. *Proceedings, Second International Symposium on Spatial Data Handling.* Williamsville, International Geographical Union: 120–1.

[195] Densham P J 1991 Spatial decision support systems. In Maguire D J, Goodchild M F, Rhind D W (eds) *Geographical information systems: principles and applications.* Harlow, Longman/New York, John Wiley & Sons Inc. Vol. 1: 403–12.

[196] Harris B 1989 Beyond geographic information systems: computers and the planning professional. *Journal of the American Planning Association* 55: 85–92.

[197] Harris B, Batty M 1993 Locational models, geographic information, and planning support systems. Journal of Planning Education and Research 12: 184–98.

[198] Batty M 1995 Planning support systems and the new logic of computation. *Regional Development Dialogue* 16: 1–17.

[199] Klosterman R E 1995 Planning support systems. In Wyatt R, Hossain H (eds) Proceedings, *Fourth International Conference on Computers in Urban Planning and Urban Management Melbourne*, July Vol. 1: 19–35.

[200] Worrall L (ed.) 1990 Geographic information systems: developments and applications. London, Bellhaven Press Yeh A G-O 1990 A land information system in the programming and monitoring of new town development. *Environment and Planning B: Planning and Design* 17: 375–84.

[201] Jankowski P 1995 Integrating geographical information systems and multiple criteria decision-making methods. *International Journal of Geographical Information Systems* 9: 251–73.

[202] Batty M, Xie Y 1994a Modelling inside GIS: part I. Model structures, exploratory spatial data analysis and aggregation. *International Journal of Geographical Information Systems* 8: 291–307.

[203] Batty M, Xie Y 1994b Modelling inside GIS: part II. Selecting and calibrating urban models using ARC/INFO. *International Journal of Geographical Information Systems* 8: 451–70.

[204] Batty M 1990 Information systems and technology for urban and regional planning in developing countries: a review of UNCRD's research project, Vol. 1. Nagoya, Japan, United Nations Centre for Regional Development.

[205] Batty M 1992 Urban modelling in computer-graphic and geographic information system environments. *Environment and Planning B: Planning and Design* 19: 663–88.

[206] Batty M 1994 Using GIS for visual simulation modelling. GIS World 7(10): 46–8] and in relation to the development of desktop GIS by Faust N.L.

[207] Faust N L 1995 The virtual reality of GIS. *Environment and Planning B: Planning and Design* 22: 257–68.

[207] Coyne R D, Sudweeks F, Haynes D 1996 Who needs the Internet? Computer-mediated communication in design firms. *Environment and Planning B: Planning and Design* 23: 749–70.

[208] Densham P J 1994 Integrating GIS and spatial modelling: visual interactive modelling and location selection. *Geographical Systems* 1: 203–19.

[209] Masser I, Campbell H 1991 Conditions for the effective utilisation of computers in urban planning in developing countries. *Computers, Environment and Urban Systems* 15: 55–67.

[210] Drummond W J 1995 Extending the revolution: teaching land use planning in a GIS environment. *Journal of Planning Education and Research* 14: 280–91.

[211] Akom A A 1982 Information system for planning in Ghana. *Ekistics* 292: 73–7.

[212] Remote Sensing and GIS for Urban Growth Analysis in China, Shupeng Chen, Shan Zeng, and Chuangjle Xle, *Photogrammetric Engineering & Remote Sensing,* Vol. 66, No. 5, May 2000, pp. 593-598.

[213] Chen, Shunqing, 1999. *Urban Expansion and Land Value Increment, Unpublished PhD Dissertation, the Institute of geography*, Chinese Academy of Sciences, Beijing, 254 p. (in Chinese).

[214] Chen, Shupeng (chief-editor), 1986. *Tianjin Atlas of Environment Quality*, Science Press, Beijing, 270 p.

[215] Yang, Da, 1996. Present state of the overall layout of China's cities and its future outlook, *Science and Technology Review*, 6:59-62 (in Chinese.

[216] Fu, Suxing, and Guifa Cao (chief-editor), 1989. Jing-Jin-Tang *Atlas of Ecological Environment,* Science Press, Beijing, 316 p.

[216] Chen, Shupeng, Chenghu Zhou, and Shan Zeng, 1998. China's urbanization problems and pressing tasks of urban GIS, Academic Forum for Urban Geographic Information Systems, (Lin Hui, editor), Joint Laboratory of Geo-informatics, Hong Kong, 102 p. (in Chinese).

[217] Gong, Peng (chief-editor), 1997. Urban Geographic Information Systems: Methods and Applications, The Association of Chinese Professionals in GIS (Abroad), Berkeley, California, 210 p. (in Chinese).

[218] Song, Xiaodong, and Anthony Gar-On Yeh, 1995. Geographic Information System and Its Application in Urban Planning and Urban Management, Science and Technology Press, Beiing, 160 p. (in Chinese).

[219] Dai, Changda, Lingli Tang, Ping Jiang and Jianbo Lin, 1996. Satellite remote sensing for monitoring urban expansion and environmental remediation, Population Dynamics and Management of Urban Environment, The 47th MF Congress, Beijing, pp. 9-16.

[220] S. Adia, A. Rabiu, "Change Detection of Vegetation Cover, Using Multi-Temporal Remote Sensing Data and GIS Techniques," Geospatial Application Papers, Environment, 2003.

[221] Gu, Chaolin, 1997. Research on Beijing landuse/landcover change mechanism, *Journal of Nature and Resources,* 14(3):105-115 (in Chinese).

[222] Sun, Yuanzong, and Jianhua Zhou, 1994. Urban Remote Sensing, Shanghai Technology Literature Press, Shanghai, 180 p. (in Chinese).

[223] Mei, Anxing, 1992. Multi-temporal aerial remote sensing dynamic research on the zonal urbanization process of Zhenru town, Shanghai city, *Journal of East China Normal University, Special Issue of Remote Sensing,* pp. 84-92 (in Chinese).

[224] Yeh, Anthony Gar-On, and Xia Li, 1997. An integrated remote sensing and GIs approach in the monitoring and evaluation of rapid urban growth for sustainable development in Pearl River delta, China, *International Planning Studies*, 2(2):193-209.

[225] Yeh and Li, 1996. Urban growth management in Pearl River delta: an integrated remote sensing and GIs approach, *ITC Journal,* 1:77-86.

[226] Li, Xia, 1997. Integration of Remote Sensing and Geographic Information System for monitoring and evaluating agricultural land loss: A case study of Dongguan, *Acta Geographica Sinica,* 52(3):279-287 (in Chinese).

[227] Dai, Feng, Congzhou Jiang, Shunqing Chen, and Jianwei Ding, 1997. Urban planning and information technology-development and application in Guangzhou, China, presented at *the International Symposium of "Cities Towards the 21st Century,"* Beijing.

[228] Huang, Wei, 1997. Guangzhou City Large-Scale Digital Orthophoto Generation and Its Application to Urban Planning, Urban Integration Remote Sensing and Mapping Center, Ministry of Construction, Urban Planning Bureau of Guangzhou City, Guanzhou, 410 p. (in Chinese).

[229] Chen, Yu, and Jinshi Wei (chief-editor), 1999. *The Atlas of Sustainable Development in Yangtze Economic Zone,* Science Press, Beijing, 248 p. (in Chinese).

[230] Application of Remote Sensing and GIS Technique for Efficient Urban Planning in India, Ravindra Kumar Verma1*, Sangeeta Kumari2, nand R. K. Tiwary3.

[231] Javed A., Pandey S. (2004), "Land-use/land-cover analysis for waste disposal," GIS@development, http://www.gisdevelopment.net/maganize/years/2004/jun/lanuse.as, accused 6-March-2007.

[232] Impervious Surface Area and Residential Housing Density: A Satellite Perspective Eric E. Hebble, Toby N. Carlson and Ken Daniel Department of Meteorology The Pennsylvania State University, University Park, PA 16802, U.S.A. http://www.geocarto.com.hk/cgi-bin/pages1/mar01/impervious.pdf.

[233] Kaur Amarjeet, Dutta Venkatesh, Chaddha D. K. (2003), "Application of Remote Sensing and GIS tools in delineating Environmentally-Fragile-Areas (EFAs) for

sustainable land use planning: A case study of Delhi Region," Proceedings of the Gap India Conferences.

[234] Kshama Gupta, Sadhana Jain, (2005), "Enhanced capabilities of IRS P6 LISS IV sensor for urban mapping," *J. Curr. Sci,* Vol. 89, No.11, pp. 1805-1812.

[235] Remote Sensing Image Interpretation Study Serving Urban Planning Based On Gis Qian Wanga, B, C, Jianping Chena, B, C, Yi Tiana, B, C A Department Of Global Science And Resource, CUGB, Xueyuan Road, Beijing - (Wangqian12031)@163.Com B Institute Of Land Resources And High Techniques , CUGB, Xueyuan Road, Beijing - (Wangqian12031)@163.Com C Laboratory Of Beijing Land Resources Information Development, CUGB, Xueyuan Road, Beijing - (Wangqian12031)@163.Com http://www.isprs.org/proceedings/XXXVII/congress/4_pdf/80.pdf.

[236] Navalgund, R. R., Jayaraman, V., Roy, P.S. (2007), "Remote sensing applications: An overview," *J. Curr. Sci.,* Vol. 93, No.12, pp. 1747-1766.

[237] Patkar, V.N. (2003), *"Directions for GIS in Urban Planning,"* GIS@development, http://www.gisdevelopment.net/application/urban/overview/urban0042pf/htm, Map Asia Conferences, Urban Planning.

[238] Patkar, V.N. (1994), *"GIS Applications and Innovative Practices for Solid Waste Management,"* Nagarlok, Vol. 26, No. 4, pp. 25-34.

[239] Pathan, S.K., Patel, J.G., Bhanderi, R.J., Arya, A.S., Navalgund, R.R., Shimpi, P.B., Dhawale, A.K. and Landge, S.D. (2000), "Remote Sensing and GIS Based Inputs for the Preparation of a Development Plan of Pimpari-Chinchwad Municipal Corporation (PCMC) Area for the Year 2018," Proceedings of Geomatics 2000: Conference on Geomatics in Electronic Governance, 21-22 January, 2000, C-DAC, Pune, pp. UP22-UP33.

[240] Rahman Atiqure, (2006), "Application of Remote Sensing and GIS Technique for Urban Environment Management and Development of Delhi, India," Applied Remote Sensing for Urban Planning Governance and Sustainability, http://www.springerlink.com/index/x5w74277j3I13959pdf.

[241] Sudhir. H. S., Ramachandra, T.V., and Jagadish, K.S. (2003), "Urban Sprawl pattern recognition and modelling using GIS," GIS@development, http://www.gisdevelopment.net/application/urban/sprawl/mi03142.htm-11k, Map Asia conferences 2003, Urban Planning.

[241] Sudhir. H. S., Ramachandra, T.V., and Jagadish, K.S. (2003), "Urban Sprawl pattern recognition and modelling using GIS," GIS@development, http://www.gisdevelopment.net/application/urban/sprawl/mi03142.htm-11k, Map Asia conferences 2003, Urban Planning.

[242] Sudhir. H. S., Ramachandra, T.V., and Jagadish, K.S. (2003), "Urban Sprawl pattern recognition and modelling using GIS," GIS@development, http://www.gisdevelopment.net/application/urban/sprawl/mi03142.htm-11k, Map Asia conferences 2003, Urban Planning.

[243] Tiwary, D. P. (2003), "Remote Sensing and GIS for efficient Urban planning in India," GIS@development.net, http://www.gisdevelopment.net/application/urban/oveview/ma03 224.htm, Map Asia conferences 2003, Urban Planning.

[244] Organization Theory And Methodology Michael C. Jensen http://webkuliah.unimedia.ac.id/ebook/files/jensen(1983)%20organization%20theory.pdf.

[245] Land Use/Land Cover and NDVI Analysis for Monitoring the Health of Micro-watersheds of Sarada River Basin, Visakhapatnam District, India.

[246] Nooka Ratnam Kinthada, Murali Krishna Gurram, Amminedu Eadara and Venkateswara Rao Velagala Journal of Geology & Geosciences, ISSN: 2329-, 6755 February 21, 2014.

[247] Application of Remote Sensing And GIS Technique For Efficient Urban Planning In India Ravindra Kumar Vermal, Sangeeta Kumari and R. K. Tiwary, http://www.csre.iitb.ac.in/~csre/conf/wp-content/uploads/fullpapers/OS4/OS4_13.pdf.

[248] AGD Landscape & Environment 3 (2) 2009. 71-86. 71, Land Use Change Detection along the Pravara River Basin in Maharashtra, Using Remote Sensing and GIS Techniques. Veena U. Joshi – Vikas Nagare Dept. Of Geography, University of Pune, Pune – 411 007.Email: vujoshi@unipune.ernet.in, http://landscape.geo.klte.hu/pdf/agd/2009/2009v3is2_2.pdf.

[249] Kuldeep Pareta, Assesment of Carbon Stock in Shifting Cultivation Land a Case Study of Nagaland State of India, Esri International User Conference, 2013 http://proceedings. esri.com/library/userconf/proc13/papers/337_9.pdf.

[250] Faisal I. Khan*, Tahir Husain, Ramzi Hejazi, An overview and analysis of site remediation technologies, *Journal of Environmental Management* 71 (2004) 95–122, pp 96-122.

[251] Usharani Va, Sarala, Pb, Poonguzali Jc and Sharmila G, Role of remote sensing and geographic information system in natural resources management INT J CURR SCI 2012, 188-191Short Communication, ISSN 2250-1770, http://www.currentsciencejournal.info/issuespdf/USharani-Remote%20sensing.pdf.

[252] P. S. Uttarwar, Monitoring and Implementation of the Master Plan Policies in Delhi, ITPI Journal, http://itpi.org.in/pdfs/jan2004/chapter5.pdf, I: I (2004) 26-33, pp. 26-33.

[253] Kit Wong, Xiu-Rong Ren, Yang-Zhong Huang, Yi Xie, Guofa Liu, Harumi Saito, Hao Tang, Leng Wen,1 Susann M. Brady-Kalnay, Lin Mei, Jane Y. Wu, Wen-Cheng Xiong, and Yi Rao, Signal Transduction in Neuronal Migration: Roles of GTPase Activating Proteins and the Small GTPase Cdc42 in the Slit-Robo Pathway, Cell, Vol. 107, 209–221, October 19, 2001, Copyright, 2001 by Cell Press, http://www.bio.pku.edu.cn/lab/raolab/papers/Wong%20et%20al.%20Cell.pdf.

[254] David M. Theobald, Landscape Patterns of Exurban Growth in the USA from 1980 to 2020, Copyright © 2005 by the author(s). Published here under license by the Resilience Alliance. Theobald, D. 2005. Landscape patterns of exurban growth in the USA from 1980 to 2020. Ecology and Society 10(1): 32. [online] URL: http://www.ecologyandsociety.org/vol10/iss1/art32/.

[255] H.S. Sudhira, T.V. Ramachandra, K.S.Jagadish, Urban sprawl: metrics, dynamics and modelling using GIS, International Journal of Applied Earth Observation and Geoinformation 5 (2004) 29–39, file:///D:/My%20Documents/Downloads/urban_sprawl%20(1).pdf.

[256] Barnes, S-J. and Lightfoot, P.C. (2005). Formation of magmatic nickel-sulfide oredeposits and processes affecting their copper and platinum-group element contents. In Hedenquist, J.W., Thompson, J.F.H., Goldfarb, R.J. and Richards, J.P. (eds.) Economic Geology 100th Anniversary Volume, p. 179-213.

[257] Jeanne Epsteln, Karen Payne, and Elizabeth Kramer, Techniques for Mapping Suburban Spraw, Photogrammetric Engineering & Remote Sensing, Vol. 63, No. 9,

September 2002, pp. 913-918, 0099-1112/02/6809-913$3.00/0, 2002 American Society for Photogrammetry, and Remote Sensing http://asprs.org/a/publications/pers/2002journal/september/2002_sep_913-918.pdf.

[258] Jyotishman Deka, Om Prakash Tripathi, Mohamed Latif Khan, Urban growth trend analysis using Shannon Entropy approach – A case study in North-East India, International Journal Of Geomatics And Geosciences, Volume 2, No 4, 2012, © Copyright 2010 All rights reserved Integrated Publishing services, Research article ISSN 0976 – 4380, http://www.ipublishing.co.in/jggsvol1no12010/voltwo/EIJGGS3088.pdf.

[259] Michael Oloyede Alabi, Urban Sprawl, Pattern and Measurement in Lokoja, Nigeria, http://www.um.ase.ro/No13/10.pdf.

[260] "Under Western Eyes" Revisited: Feminist Solidarity through Anticapitalist Struggles, Chandra Talpade Mohanty, http://www2.hawaii.edu/~dasgupta/Mohantysigns.pdf.

[261] S.N. Miller, D.J. Semmens, R.C. Miller, M. Hernandez, D.C. Goodrich, W.P. Miller, W.G. Kepner, D.W. Ebert, GIS-Based Hydrologic Modeling: The Automated Geospatial Watershed Assessment Tool, 1USDA-ARS Southwest Watershed Research Center, 2000, http://www.tucson.ars.ag.gov/agwa/docs/posters/agwa_overview.pdf.

[263] Praveen Kumar Rai and V. K. Kumra, Role of Geoinformatics In Urban Planning, Journal of Scientific Research Vol. 55, 2011 : 11-24, Banaras Hindu University, Varanasi ISSN : 0447-9483, pp 11-24.

[264] Patkar, V.N. (2003), "Directions for GIS in Urban Planning," GIS@development, http://www.gisdevelopment.net/application/urban/overview/urban0042pf/htm, Map Asia Conferences, Urban Planning.

[265] Eric E. Hebble, Toby N. Carlson and Ken Daniel, Impervious Surface Area and Residential Housing Density: A Satellite Perspective, Geocarto International, Vol. 16, No. 1, http://www.geocarto.com.hk/cgi-bin/pages1/mar01/impervious.pdfMarch 2001, pp. 15-20.

[266] A.P. Krishna and Suman Kumar Mitra, Geoinformatics Based Environmental Quality Assessment of Physical Parameters of Urbanization: Case Study of Ranchi City, India, file:///D:/My%20Documents/Downloads/0f31753500fd94120c000000.pdf.

[267] National Natural Resources Management System (NNRMS) (2004). National Urban Information System (NUIS) Project Document. Indian Space Research Organization and Ministry of Urban Development, Government of India, p. 77.

[268] Anne Vernez Moudon and Michael Hubner (2004) Monitoring Land Supply with Geographic Information Systems, John Wiley and Sons Publication, New York.

[269] Dhawale, A.K. and Landge, S.D. (2000), "Remote Sensing and GIS Based Inputs for the Preparation of a Development Plan of Pimpari-Chinchwad Municipal Corporation (PCMC) Area for the Year 2018," Proceedings of Geomatics 2000 : Conference on Geomatics in Electronic Governance, 21-22 January, 2000, C-DAC, Pune, pp. 22-33.

[270] John Stillwell, Stan Geertman and Stan Openshaw (eds.) Geographic Information and Planning, Springer Publication, New York.

[271] Arup Kumar Sarma, Lecture Notes for Workshop on "Integrated Planning for Hazard Mitigation in Hilly Urban Area" 21st -22nd January 2010 at Centre of Excellence for Integrated Landuse Planning and Water Resource Management (ILPWRM), Department of Civil Engineering Indian Institute of Technology Guwahati Guwahati - 781 039.

[272] Jean-Paul Donnay, Michael J. Bransley and Paul A. Longley (eds.) Remote Sensing and Urban Analysis, Taylor and Francis Publications, London.
[273] Tiwari, D.P. (2006), Remote Sensing and GIS for efficient Urban Planning, GIS Development.
[274] Victor Mesev (ed.) Remotely Sensed Cities, Taylor & Francis Publications, London and New York.
[275] Ebenezer Howard (1898), Garden City Movement, http://en.wikipedia.org/wiki/Garden_city_movement.
[276] M.J. Fasona and A.S. Omojola, GIS and Remote Sensing For Urban Planning: A Case Of Festac Town, Lagos, Nigeria, Geoinformatics 2004, Proc. 12th Int. Conf. on Geoinformatics − Geospatial Information Research: Bridging the Pacific and Atlantic University of Gävle, Sweden, 7-9 June 2004.
[277] Brockerhoff, M.P., 2000: An urbanising world. Population Bulletin, Vol. 55, No. 3, Population Reference Bureau, 45 p.
[278] Oyeleye, D.A., 2001: Settlement Geography, University of Lagos Press.
[279] Selman, P., 1996: Local Sustainability – Managing and Planning Ecologically Sound Places. Paul Chapman Publishing Ltd., London.
[280] Mahesh Jhawar, Nitin Tyagi, Vivek Dasgupta, Urban Planning Using Remote Sensing, IJIRSET 2319 – 8753, International Journal of Innovative Research in Science, Engineering and Technology, Vol. 1, Issue 1, November 2012, Copyright to IJIRSET www.ijirset.com 42.
[281] Sudhir. H. S., Ramachandra, T.V., and Jagadish, K.S., Urban sprawl pattern analysis using GIS, Bangalore, Centre for Ecological Sciences, CES Technical Report N 99, (2003).
[283] Urban Sprawl pattern recognition and modelling using GIS," GIS@development, http://www.gisdevelopment.net/application/urban/sprawl/mi03142.htm-11k, Map Asia conferences 2003, Urban Planning.
[284] Assefa M. Melesse, Spatiotemporal dynamics of land surface parameters in the Red River of thc North Basin, Physics and Chemistry of the Earth 29 (2004), pp. 795–810.
[285] Tiwary, D. P. (2003), "Remote Sensing and GIS for efficient Urban planning in India."
[286] GIS@development.net, http://www.gisdevelopment.net/application/urban/oveview/ma03 224.htm, Map Asia conferences 2003, Urban Planning.
[287] http://www.satimagingcorp.com/galleryimages/ikonos-new-delhi-taj-mahal.jpg; http://media.wiley.com/product_data/excerpt/8X/04707495/047074958X-192.pdf; http://www.satimagingcorp.com/characterization-of-satellite-remote-sensing-systems.html; http://www.eoearth.org/article/Remote_sensing; http://www.earthzine.org/2011/07/20/remote-sensing-%E2%80%93-an-effective-data-source-for-urban-monitoring/.
[288] http://www.sensysmag.com/article/features/13931-remote-sensing-and-urban-planning-a-common-future-.html].
[289] http://ieeexplore.ieee.org/xpl/articleDetails.jsp?reload=true&arnumber=1219987 &contentType=Conference+Publications;
[290] http://www.isprs.org/proceedings/XXXVI/8-W27/netzband.pdf.
[291] Nnam, Victor CHUKWUEMEKA, Nnam, Godwin Uchechukwu and Nnam John OKWOR, Nigeria, Assessment of Urban Development Planning using Supervised Classification of remotely sensed Imageries and GIS, A Case study of Independence

Layout (Part of), Enugu, Nigeria, TS03B - Remote Sensing for Landuse and Planning – 6388, FIG Working Week 2013

[292] Burrough, P. A. and McDonnell, R. A. (1998). Principles of Geographical Information Systems, Oxford University Press, New York.

[293] Richards, J. A., *Remote Sensing digital image analysis: an introduction* (second edition), 1993 p. 85.

[294] Jensen, J.R., (1996), *Introductory digital image processing: a remote sensing perspective,* Upper Saddle River, NJ: Prentice Hall, (second edition).

[295] Nnam U. G (2012). Comparison of Pixel based and Object based Image Classification for Mapping Urban Greenery in Uwani, Enugu. DPR 762. Department of Geoinformatics and Surveying, University of Nigeria, Enugu Campus.

[296] Carleer, A. P. and Wolff, E., 2006. Urban land cover multi-level region-based classification of VHR data by selecting relevant features. *International Journal of Remote Sensing,* 27(6), pp. 1035–1051.

[297] Carleer, A.P., Debeir, O. and Wolff, E., 2005. Assessment of very high spatial resolution satellite image segmentations. *Photogrammetric Engineering and Remote Sensing,* 71, pp. 1285-1294.

[298] Chen Y., Shi P., Fung T., Wang J., Li Y., (2007). Object-oriented classification for urban land cover mapping with ASTER imagery. *International Journal of Remote Sensing* 28 (29), 4645_4651.

[299] Luyao Huang and Ling Ni (2008), Object-Oriented Classification of High Resolution Satellite Image for Better Accuracy, *Spatial Accuracy Assessment in Natural Resources and Environmental Sciences,* June 25-27, pp. 211-218.

[300] Nargiz A. Babayeva, Rayxa B. Amenzade, Rustam B. Rustamov, *Space Technology Advances Application In Town Planning.*

[301] Wentz, Elizabeth A., David Nelson, Atiqur Rahman, William L. Stefanov, Shoursaseni Sen Roy 2008. Expert system classification of urban land use/cover for Delhi, India *International Journal of Remote Sensing 29* (15): 4405-4427.

[302] Keys, Eric, Elizabeth A. Wentz, Charles Redman 2007. The spatial structure of land use from 1970-2000 in the Phoenix, Arizona metropolitan area. *Professional Geographer* 59(1): 131-147.

[303] Myint, S.W., Wang, L. 2006. Multi-criteria Decision Approach for Land Use Land Cover Change Using Markov Chain Analysis and Cellular Automata Approach. *Canadian Journal of Remote Sensing* 32(6):390-404.

[304] Myint, S.W., Jain, J., Lukinbeal, C., Lara-Valencia, F., 2010. Simulating urban growth on the U.S.-Mexico border: Nogales, Arizona and Nogales, Sonora, *Canadian Journal of Remote Sensing,* 36(3): 166-184.

[305] Praveen Kumar Rai and V. K. Kumra, Role Of Geoinformatics In Urban Planning, Department of Geography, B.H.U., Varanasi-221005, Journal of Scientific Research Vol. 55, 2011 : 11-24, Banaras Hindu University, Varanasi ISSN : 0447-9483.

[306] Nargiz A. Babayeva [1] Rustam B. Rustamov [2], Remote Sensing and GIS Applications in Architectural Design.

[307] Sefer Kurnaz, Rustam B. Rustamov. Application of Space Technology in Support of Security and Safety of Critical Infrastructure, H. Gonca Coskun et al. (eds.), Integration of Information for Environmental Security, 149-154, @ 2008 Springer, The Netherlands.

[308] Sefer Kurnaz, Rustam B. Rustamov, Maral Zeynalova and Fazila Qasimova. Natural growth as an indicator of monitoring of the linear infrastructure for safety and security issues. *KSAS International Journal* (The Korean Society for Aeronautical & Space Sciences), May 2008, v.9 № 1, p.59.

[309] Maral H. Zeynalova, Rustam B. Rustamov, Saida E. Salahova, "Space Technologies for the Benefit of Human Society and Earth," P. Olla (ed.), DOI 10.1007/978-I-4020-9573-3_5, © Springer Science+Business Media B.V. 2009.

[310] Sefer Kurnaz, Gunel Babayeva, Rustam B. Rustamov, Elman Aleskerov, Monitoring of the Linear Infrastructure: Environmental and Social Impacts, Proceedings of 4th *International Conference on Recent Advances in Space Technologies, IEEE,* Istanbul, Turkey, June 11-13, 2009, pp. 273-276.

[311] N.A. Babayeva, R.B. Amanzade, N. Rahimov, R.B. Rustamov, Space Technology Advances Application in Town Planning, *Journal of Environmental Science and Engineering Technology*, Volume 2, Issue1, pp. 9-13.

[312] UNESCO World Heritage: Gobustan Rock Art Cultural Landscape.

[313] Shannon Jackson, GIS as an Urban Planning Tool, *The International Journal of Technology,* Knowledge and Society, Volume 2, Issue 7, pp.107-116.

[314] Sefer Kurnaz, Rustam B. Rustamov, Maral Zeynalova and Fazila Qasimova. Natural growth as an indicator of monitoring of the linear infrastructure for safety and security issues. *KSAS International Journal* (The Korean Society for Aeronautical & Space Sciences), May 2008, v.9 № 1, p.59.

[315] Maral H. Zeynalova, Rustam B. Rustamov, Saida E. Salahova, "Space Technologies for the Benefit of Human Society and Earth," P. Olla (ed.), DOI 10.1007/978-I-4020-9573-3_5, © Springer Science+Business Media B.V. 2009.

[316] Sefer Kurnaz, Gunel Babayeva, Rustam B. Rustamov, Elman Aleskerov, Monitoring of the Linear Infrastructure: Environmental and Social Impacts, *Proceedings of 4th International Conference on Recent Advances in Space Technologies, IEEE,* Istanbul, Turkey, June 11-13, 2009, pp. 273-276.

[317] "State Oil Company of Azerbaijan Republic: Transition from National to Transnational Company or Demand of Time?." Turkishweekly.net. 28 January 2011. Retrieved 17 December 2011.

[318] "Population by economic and administrative regions, urban settlements at the beginning of the 2009." Archived from the original on 14 November 2009. Retrieved 21 November 2009.

[319] "Port of Baku." *World Port Source.* Retrieved 23 December 2010.

[320] "Azerbaijani FM: Baku-Tbilisi-Kars railroad to be built in 2012." trend.az. Retrieved 26 January 2011.

[321] Railway Gazette International February 2009 p. 54 with map.

[322] "Baku Sea Port expects for passenger trip Anzali-Baku." En.trend.az. 18 February 2010. Retrieved 17 April 2010.

[323] "Ferry Service across Caspian sea." Azerbaijan24.com.Archived from the original on 29 March 2010. Retrieved 17 April 2010.

[324] "Baku Sea Port." Bakuseaport.az. Archived from the original on 16 April 2008. Retrieved 7 March 2008.

[325] "Azerbaijan, Baku." m-hesse.com. Retrieved 1 February 2011.

[326] "International Airport (BAK/UBBB), Baku, Azerbaijan." Airport-technology.com. Archived from the original on 12 March 2010. Retrieved 17 April 2010.

[327] The Day of Military Air force of Azerbaijan, http://en.wikipedia.org/wiki/Azerbaijani_Armed_Forces.

[328] "Administrative, density and territorial units and land size by economic regions of Azerbaijan Republic for January 1. 2007." Archived from the original on 24 November 2007. Retrieved 17 July 2009.

[329] The State Statistical Committee of the Republic of Azerbaijan, 2.5 Population by sex, economic and administrative regions, urban settlements at the beginning of 2012, retrieved on October 2, 2012.

[330] "Travel Picks: Top 10 cities to party the night away." Reuters. 13 November 2009. Archived from the original on 23 November 2009. Retrieved 21 November 2009.

[331] A. Hashimov, B. N. Aliyeva, S.R. Rustamova, R. B. Rustamov, An Environmental Impact Assessment With Use of Space Technology Advances, *The Tenth International Conference on Technical and Physical Problems of Electrical Engineering,* ISSN 2309-0545 and ISSN 2309-0553, Baku, Azerbaijan, September 07-08, 2014, p. 357-360.

[332] T. Mashally, *"Green Economics or Projects of Many Billions?"* http://www.novoye-vremya.com, Baku, Azerbaijan, 2014.

[333] "Satellite Monitoring of the Southeastern Baltic Sea," *Annual Report* 2004 of LUKOIL.

[334] Schnick S and Tao V., "Application of LIDAR Technology for Pipeline Mapping and Safety, Proceeding of ISPRS WG III2," Workshop on Three-dimensional Mapping from InSAR and LIDAR, 2001.

[335] V. K. Bansal and Mahesh Pal, *"Geographic Information Systems in Evaluation and Visualization of Construction Schedule,"* Second ESRI Asia-Pacific User Conference New Delhi, 2007.

[336] T. Vanderstraete, R. Goossens, and T.K. Ghabour, "Realization of a Local Coral Reef Monitoring System in the Red Sea using Landsat7 ETM+ data," In *Proceedings of the 24th EARSeL Symposium 'New Strategies for European Remote Sensing,'* Ed. M. Oluic, Dubrovnik (Croatia), 25 – 27 May 2004, pp. 479-488, 2005.

[337] Zhong, D., Li, J., Zhu, H., and Song, L., ''Geographic information system based visual simulation methodology and its application in concrete dam construction processes.'' *J. Constr. Engrg. and Mgmt., ASCE,* 130(5), 742–750, 2004.

[338] *Geotechnical Investigation Report for Different Areas of Absheron Peninsula,* Encotec Company, Baku, Azerbaijan, 2012.

[339] *Chemical Analysis of Different Soils of Absheron Peninsula,* Encotec Company, Baku, Azerbaijan, 2014.

[338] B. Aliyeva, R.B. Rustamov, Remote Sensing Methods and GIS Technology Applications in Land Use Purposes for Engineering Solutions, *The Ninth International Conference on Technical and Physical Problems of Power Engineering,* ISSN 2309-0545 and ISSN 2309-0553, Istanbul, Turley, September 09-11, 2013, p. 426-429.

[339] B.N. Aliyeva, R.B. Rustamov, Integration Space And Field Data In Land Use Studies For Engineering Solutions, Issue 17, Volume 5, Number 4, December 2013 (Serial No: 0017-0504-1213) of *International Journal on "Technical and Physical Problems of Engineering" (IJTPE),* pp. 130-133.

[340] Giles D., 1994, "A geographical information system for geotechnical and ground investigation data and analysis." *Fifth European Conference Exhibition of Geographic Information System,* Vol. I, pp. 776-778.

[341] Neteler M., 1998, Introduction to GRASS GIS software. Hanover, Germany, available at http://www.geog.uni-hannover.de.

[342] Andrus, R., Stokoe, K., Chung, R., and Juang, C., (2003), "Guidelines for Evaluating Liquefaction Resistance using Shear Wave Velocity Measurement and Simplified Procedures," Report No. NIST GCR 03-854, National Institute of Standards and Technology, Gaithersburg, MD, 151 pp.

[343] Deaton, S.L., Frost, J.D., Luna, R., and Parsons, R.L., (2001), "GIS-Based Evaluation of Geotechnical Borehole Log Quality," *Journal of the Transportation Research Board,* TRR No. 1755, TRB/NRC, Washington, D.C., pp. 15-25.

[344] McGraw-Hill "Geotechnical Engineering's portable handbook," ISBN 0-07-135111-6, 2000.

[345] NAVFAC Design manual 5.4 "Civil Engineering," 1979.

[346] Deaton, S.L., and Frost, J.D., (2002), "Integrated Digital Earthquake Reconnaissance," 7th U.S. National Conference on Earthquake Engineering, CD-ROM.

[347] A.M. Hashimov, R.B. Rustamov, S.N. Hasanova, S.R. Rustamova, Space Technology Applications in Power Energy Environmental Problems, *The Eight International Conference on Technical and Physical Problems of Power Engineering,* ISSN 2303-9752 and ISSN 2303-9760, Norway, Fredrikstad, September 5-7, 2012, p. 442-444.

[348] S.N. Hasanova, R.B. Rustamov, Remote Sensing Space Technology as a Tool of Environmental Problems in Power Energy Transmissions, *The Ninth International Conference on Technical and Physical Problems of Power Engineering,* ISSN 2309-0545 and ISSN 2309-0553, Istanbul, Turley, September 09-11, 2013, p. 441-444.

[349] Eric P. Crist, *Short Communication a TM Tasseled Cap Equivalent Transformation for Reflectance Factor Data,* OElrwier Science lblirhing Ca... Inc., 1985, 52 Vanderbilt Ave., New York. NY 10017, pp. 301-306.

[350] Crist, E. P. and Kauth, R. J., T*he Tasseled CapDeMy-Stified, Photogrammetric Engineering and Remote Sensing,* Vol. 52, No. 1, January 1986, pp. 81-86.

[351] Lilesand, T.M. and R.W. Kiefer, Review of "Remote Sensing and Image Interpretation, *International Journal of Remote Sensing* Volume 8, Issue 12, 1987, pp. 721.

[352] A Hashimov, S.E. Salahova, M.H. Zeynalova, R.B. Rustamov, Human Impact of Environmental Problem Solving by Advances of Space Technology, *The Ninth International Conference on Technical and Physical Problems of Power Engineering,* ISSN 2309-0545 and ISSN 2309-0553, Istanbul, Turley, September 09-11, 2013, p. 445-447.

[353] Bern, T-I., Wahl, T. Andersen, and R. Olsen. "Oil Spill Detection Using Satellite Based Sar: Experience From A Field Experiment," *Photogrammetric Engineering And Remote Sensing,* 59(3), pp. 423-428, 1993.

[354] Dahdouh-Guebas, F., Kairo, J.G., Jayatissa, L.P., Cannicci, S. and Koedam, N. 2002, 162.

[355] Dahdouh-Guebas, F., N.: 2002, 4(2), William E. Roper, USA.

AUTHORS' CONTACT INFORMATION

Dr. Rustam B. Rustamov
Institute of Physics,
Azerbaijan National Acadamy of Sciences, Senior Scientist
22 A. Javad 91, AZ1001, Baku, Azerbaijan
Email: r_rustamov@hotmail.com

Bahar N. Aliyeva
Engineering Consulting Technologies LLC,
Baku, Azerbaijan

Nargiz A. Babayeva
Engineering Consulting Technologies LLC,
Baku, Azerbaijan

Sabina N. Hasanova
British Petroleum,
Baku, Azerbaijan

Maral H. Zeynalova
Institute of Botanics Azerbaijan
National Academy of Sciences,
Baku, Azerbaijan

INDEX

#

A

B

C

D

E

F

I

J

K

L

M

N

O

P

Q

R

S

T

U

V

W

Y

Z